Also by Raymond M. Smullyan

THIS BOOK NEEDS NO TITLE

A Budget of Living Paradoxes

by Raymond M. Smullyan

A TOUCHSTONE BOOK
Published by Simon & Schuster, Inc.
NEW YORK

Copyright © 1980 by Raymond M. Smullyan

First Touchstone Edition, 1986

Published by Simon & Schuster, Inc.
Simon & Schuster Building
Rockefeller Center
1230 Avenue of the Americas
New York, New York 10020

Originally published by Prentice-Hall, Inc.

TOUCHSTONE and colophon are registered trademarks
of Simon & Schuster, Inc.

Manufactured in the United States of America

10 9 8 7 6 5 4 3 2   Pbk.

Library of Congress Cataloging-in-Publication Data

Smullyan, Raymond M.
  This book needs no title.

  (A Touchstone book)
  Reprint. Originally published: Englewood Cliffs,
N.J.: Prentice-Hall, c1980.
  1. Paradoxes. 2. Philosophical recreations.
I. Title.
[PN6361.S6 1986]   793.73   86-17734
ISBN 0-671-62831-3 Pbk.

Contents

I FABLES AND FANCIES

Should One Worry?

☐ If one worries a lot, one is obviously unhappy, since worry itself is one of the most painful things in life. If one fails to worry enough, then (at least so I have been told) one may be even worse off because one may fail to take the precautions necessary to ward off even greater catastrophes than worry.

Who is really better off, the happy-go-lucky who enjoys himself from day to day and lets tomorrow take care of itself, or the worrying, prudential person who takes all conceivable precautions for the future but is always worrying that he is not taking *enough* precautions?

All my life, people have told me that my main trouble is that I do not worry enough. And I must admit, this thought has always worried me.

The Man Who Did Not Want to Be Envied

☐ Once upon a time there was a man who could not stand being envied. He was a very creative individual and was a professor at one of the world's most prestigious universities. All his colleagues outside his university envied him his position. This he could not stand. So he gave up his position and deliberately chose a teaching position at a fifth-rate university. Still his creative work persisted and he published. Everyone envied him for the marvelous quality of his work, and also for his marvelous modesty in not seeking a more prestigious position. So he decided to publish anonymously. This did not fool anybody; nobody but *he* could write the way he wrote! So he then decided that he would continue his writing, but not publish any more in his lifetime. He would arrange that all his work should be published after his death. If people should envy him posthumously, how would that hurt *him*? Thus years went

3

by, and he did not publish. So he got fired. The fifth-rate university said, "You see, we want our professors to publish; we want them to *amount to something*." When he reminded them of his past publications, they said, "Yes, but what have you done for us *recently*?" Actually, the man did not very much mind being fired; it fitted in quite well with his present way of life. So he took his savings (which were quite large, since he had always lived very frugally), retired to the country, where he found an incredibly beautiful spot, and with his own hands built a small but absolutely gorgeous house. The spot he had chosen was by far the most beautiful for miles and miles and miles around. Everybody who passed it turned green with envy. Indeed, sometimes he could not tell whether he was looking at a tree or at an envious spectator. All this, of course, he could not stand. So he sold all his wordly possessions and gave every cent to the poor. Then he entered life as a wandering mendicant. For the first time in his life, he was free from being envied. For the first time in his life, he was truly happy. Until one day—years later—he met an old colleague from his prestigious university. After they had lunched and chatted together, the colleague said, "How radiantly happy you look; God, how I *envy* you!"

The Man Who Wanted a Tranquil Life

□ Once there was a man who wanted a life of complete serenity. The funny thing is that he spent all his days griping that he didn't have one.

Selfish or Unselfish?

☐ Once there was a white man who devoted all his time and energies to working for the cause of the black people—working to obtain complete economic, political, and psychological equality. One day a black friend told him: "We certainly admire you! Making every sacrifice under the sun to help us! We have seldom met anybody that devoted, that unselfish!" The man replied: "Unselfish? No, you don't understand; you don't understand at all! I don't have an unselfish bone in my body; everything I do is for an ulterior motive!

"You see, I happen to believe in reincarnation. When I die, many babies will be born, some black and some white. My chances of being black or white in my next life are roughly equal, and I want to be sure that if I come back black, I don't get hurt!"

Happiness

☐ *Jim:* Are you happy at this very moment?

John: In one way, yes; in another way, no.

Jim: Can you be more explicit?

John: Yes, I have just heard the first really convincing argument for the immortality of the soul. Now I know for sure that I will survive my bodily death. This makes me very happy. On the other hand, my steak is overdone.

Types of Optimists

☐ Do you know the difference between an optimist and an

5

incurable optimist? Well, an optimist is one who says: "Everything is for the best; mankind will survive."

An incurable optimist is one who says: "Everything is for the best; mankind will survive. And even if mankind doesn't survive, it is still for the best."

Then there is what I would call a *pessimistic* optimist. A pessimistic optimist is one who sadly shakes his head and says: "I'm very much afraid that everything is for the best!"

Story of the Egotist

☐ Once upon a time there was an extremely egocentric man. "I," "I," "I," was all he ever talked about. He did it in a very charming, witty, entertaining and delightful way—still it was always "I," "I," "I." Despite his extreme narcissism and exhibitionism, he delighted and benefited everyone he contacted—indeed everybody loved him. Still, all he could talk about was "I," "I," "I"! He was *terribly* ashamed and guilty about his own egocentricity! He kept saying, "How horrible, *I* am egocentric, *I* am egocentric, *I* am egocentric." You see, he had been taught from earliest childhood that nothing on earth was more shameful, disgraceful, and despicable than an egotist. So he was ashamed, guilty, and disgraced in his own image, and in his image of God's image of him. But all his shame and guilt were totally powerless to cure him of his egotism—indeed, for obvious reasons, it only made him worse. Then came the great day of enlightenment! He read somewhere or other that his approach to the problem of egotism was all wrong! The worst thing you can do is be *ashamed* or *guilty* for your egotism; that only *accentuates* it. The right way is not to approach the problem from a *moral* angle at all, but just to realize that the real disadvantage of being egocentric is that it makes you *suffer*. You will be much *happier* if you abandon your egocentricity. At first our hero was quite puzzled about the idea of abandoning his egotism just in order that *he* be happier—this only seemed to be egotism all over again. But he soon got over that hurdle and decided that this was his wisest course. So he emerged a changed man. When he went forth into society, it was no longer "I," "I," "I," but "you," "you," "you," "he," "he," "he," "she," "she," "she," etc. And everybody was so delighted! They all said, "How he has changed! How he has matured! He is no longer an egotist! Instead of always talking about himself, he now at last talks about *me!*" And so they all congratulated him upon his magnificent change. Unfortunately, what they did not realize was that the terrible strain of this ex-egotist in abandoning his former egotism was so great that he went into a deep depression which lasted the rest of his life.

A Bit on Egocentricity

☐ *Moralist:* Your whole trouble is that you are too egocentric. All you ever think or speak about is yourself—your problems, the ways you have been traumatized, your reactions to them, etc. You seem to have absolutely no interest in other people.

Victim: This obviously can't be so! As you well know, my main interest these days is in biographies and autobiographies. If I weren't so utterly fascinated by these people, why would I spend so much time reading and thinking about them?

Moralist: Because all these people you spend so much time reading and thinking about are people who are exactly like you!

Portrait of an Unsuccessful Egotist

☐ A couple was visiting us, and the wife complained that the husband was a "tyrant." I kindly asked the husband, "Are you a tyrant?" He sadly answered, "Ah, yes, but unfortunately not a very successful one!"

This made me realize that I might be most aptly described as an "unsuccessful egotist." Most people have been sufficiently influenced by moralists to have a high esteem for altruism and a poor esteem for egotism. Not so with *me*! I *love* egotists! My problem is not that I am *too* egocentric, but that I am not nearly egocentric enough—at least for my tastes. I wish I could be *more* egocentric, but I don't know how to go about it! I am an egotist in *principle* but fail to be one in fact. But don't feel too sorry for me—I'm working on it!

Incidentally, I'm also an unsuccessful altruist. The fact that I love egotists doesn't mean I don't also love altruists. I do love them—at least as much as I love egotists. And so I also would like to be far more altruistic than I am. I don't know how to go about that either.

Story of the Fleas

☐ Once upon a time there was a man. This man had a dog. This dog had fleas. The fleas infected the entire household. So the man had to get rid of them. At first he tried killing them individually with a fly swatter. This proved highly inefficient. Then he tried a flea swatter. This was also inefficient. Then he suddenly recalled, "There is such a thing as science! Science is efficient! With modern American equipment, I should have no trouble at all!" So he purchased a can of toxic material guaranteed to "kill all the fleas," and he sprayed the entire house. Sure enough, after three days all the fleas were dead. So he joyously exclaimed, "This flea spray is marvelous! This flea spray is *efficient!*"

But the man was wrong. The flea spray was actually totally inefficient. What really happened was this: Although the spray was inefficient, it was highly odiferous. Hence he had to open all the windows and doors to ventilate. As a result, all the cold air came in, and the poor fleas all caught cold and died.

Is Peekaboo a Machine?

☐ This story—though completely untrue—will, I hope, convince my readers of the value of mathematical analysis.

Peekaboo is one of my dogs. Is she a mere machine? If she is, it is extremely difficult to find out how she works! I have often wondered, "Just how does this remarkable machine work?" Inorganic machines like automobiles or, for that matter, even the most advanced electronic equipment are often quite complicated, but to discover their *modus operandi* is child's play compared with Peekaboo!

How *does* Peekaboo work, anyhow? This problem vexed me for a long time, until I finally found the answer! And here is where Science and Mathematics come in. Without these two disciplines, I would have remained in the dark forever. But now I know!

It happened this way. I realized that no mere armchair philosophical theorizing would solve the problem; what was needed were *experiments*. And so I hired one hundred of the world's leading experimental psychologists to keep Peekaboo under observation for several months, and to most carefully record all relevant data. Finally I had sufficient data to solve the problem completely, if only I could interpret them correctly! I finally boiled down the correct interpretation to the matter of simultaneously solving a system of 105 partial differential equations. Many of these equations are extremely long—indeed, one of them fills several volumes. The task of solving them is prodigious—even the most facile mathematician would require several lifetimes—assuming the equations could be solved at all! If I had lived twenty years ago, the solution would have been impossible to obtain. But now we have high-speed computers! So I rented the fastest computer in the world, programmed in all the information, and waited. It took several months, but at last the day arrived when the equations were solved! Now I have the whole key, and Peekaboo's behavior is no longer a mystery to me. I know now *exactly* how she works. I can predict her every action to a T. Given *any* stimulus whatsoever, I know exactly her response.

Take, for example, the question of obedience. In the past, when I gave her a command, I had absolutely no way of knowing how she would respond. Her responses seemed to me so varied that I could find no general law which would govern them all. Now I have such a law. But remember, I could never have found out this law without mathematical analysis.

What is this law? I will tell you. Every time I have ever given her a command, she has *always* responded in exactly the same way, only I was not bright enough to recognize what the way is. Unaided by mathematics, I kept looking at the *differences* of the responses and was totally blind to the similarities. But now I know! Whenever I give her a command, there is only one thing she ever does, and every time it is the same thing! Either she obeys it or she doesn't.

Story of the Suitor

☐ Once there was a man who was in love with a minister's daughter and planned to ask the minister for her hand in marriage. One evening he got invited to the house for dinner.

He spent the entire evening earnestly conversing with the minister about religious topics; he tried his best to impress him with his own theological erudition and great spirituality. After he left, the minister said to his daughter, "Dear, I think you should marry someone a little more practical."

Modesty

☐ *A:* For a person of your accomplishments, you are remarkably modest!

B: I'm not modest.

A: Ah, I've caught you! By disclaiming your modesty, you're trying to create the impression that you are so modest that you won't take credit for anything—not even your modesty! But I see through you! You are *affecting* the air of modesty, but in so doing, you are being most immodest!

B: It's like I said—I'm not modest.

Modesty?

☐ A counterpart to the preceding piece is the story of a man who had the reputation of being the world's most modest person. He signed all his letters, "He who is modest."

Well, a student was once discussing this matter with his rabbi. He said, "Now, how could he possibly be modest when his signing himself *He who is modest* clearly belies the fact?"

The rabbi replied: "You don't understand; you don't understand at all! He *is* completely modest. Modesty has so thoroughly entered his soul that he no longer regards it as a virtue."

11

Illogicalities

☐ The following stories are all true.

A reporter once gave the following account of an automobile accident involving a woman: "She was physically unhurt, but she was in such a state of shock that she was unable to confuse fantasy with reality."

A radical was once giving a speech on a soapbox in Greenwich Village and was talking about the plight of the poor artist. He excitedly said, "And a painter without paints can't paint—unless he has canvases!"

An employer once said to one of his employees, "Now, you tell me your needs and I'll see to it that it ought to be done."

A friend of mine who had not seen me for a year said, "You've hardly changed at all! It's only been a year since I saw you."

I once heard a bus driver say, "Federal regulations strictly forbid smoking, only in the last three rows."

I once saw a sign in a restaurant which read: "Special prices for schools, clubs, churches, and other occasions."

A student I know told me that when he was in high school he did very poorly in mathematics. As a result, he was sent to the principal's office, and the following dialogue ensued:
Principal (in a thunderous voice): Why are you doing badly in mathematics?
Student: I don't like mathematics!
Principal: Oh, but you've *got* to like mathematics! Just think now, suppose you graduate without knowing any mathematics. You go into a grocery store and your bill is eighty-seven

cents. You give the grocer a dollar bill, and he gives you only thirteen cents change. If you knew no mathematics, you wouldn't even know the difference!

I once saw a sign by a fruit stand which read: "Bananas—8¢ apiece, or three for a quarter."

The mathematician Professor Boas relates the following story: He and another mathematician, Dan Finkbeiner, had just finished lunch in a restaurant. The waiter came over and asked, "You want your checks separate or together?" Professor Boas replied, "Separate." The waiter then asked Professor Finkbeiner, "You want yours separate too?"

I know one philosopher who has the following remarkable characteristic: He makes a statement. You challenge the statement. He then gives a long, elaborate (and highly confused) argument which invariably ends up proving the very opposite of his original statement.

Oscar Collier of Prentice-Hall has kindly given me his permission to report the following incident.[1]

He passed a restaurant on which was the following sign:

```
SPECIALS
MONDAYS—ROAST BEEF
TUESDAYS—CLOSED
```

[1]Actually, I never asked him, but I'm sure he would, if I did.

13

II WHY IS LIFE SO RIDICULOUSLY PARADOXICAL?

Can't or Won't?

☐ I know one woman who smokes. She says: "It's not that I *have* to smoke; I *choose* to. I could easily give it up any time I wish to, but I see no reason why I should. But I can assure you, I could if I wanted to." Her husband says to her: "That's only a rationalization! You couldn't give up smoking even if you wanted to. You are not strong enough to give it up; you *have* to smoke. So to make yourself feel better, and to avoid having to confess your own weakness of character, you fool yourself into believing that you *choose* to. But it's only a rationalization!"

I know another woman who smokes. She says: "It's not that I *want* to smoke; I can't help myself! I have tried several times giving it up, but I have failed! I'm afraid I just don't have a very strong character. I would love to stop, but I simply can't." Her husband says to her: "That's only a rationalization! You certainly could stop immediately, if you really wanted to. No, you choose to smoke (after all, nobody is *making* you) and you feel ashamed and guilty for doing that which you know to be harmful. So to avoid any moral responsibility for your acts, you fool yourself into believing that you 'can't help it.' But this is only a rationalization."

My only question about all this is: "Why are people so incredibly stupid?"

A Rationalist and His Wife
Or "Why wives don't wish their husbands to be overly rational"

☐ *Wife:* Do you love me?
Rationalist: Well of course! What a ridiculous question!
Wife: You *don't* love me!
Rationalist: Now what kind of nonsense is this?

17

Wife: Because if you *really* loved me, you couldn't have done what you did!

Rationalist: I have already explained to you that the reason I did what I did was *not* that I don't love you, but because of such and such.

Wife: But this such and such is only a *rationalization*! You *really* did it because of so and so, and this so and so would never be if you really loved me.

Etc., etc.!

Next Day

Wife: Darling, do you love me?

Rationalist: I'm not so sure!

Wife: What!

Rationalist: I thought I did, but the argument you gave me yesterday proving that I don't is not too bad!

On Shutting Up!

☐ I have heard much these days about "scream therapy." People come together in a group and are encouraged to "scream at" each other. It is said to be good to "scream one's hostility out of one's system."

This may not be a bad idea! But perhaps we should also have group "shut up" therapy sessions. That is, people come together and learn to "shut up" at each other. I say "shut up *at*" advisedly, because people sometimes do shut up *at* each other! When we are angry, we scream at each other, and when we are still angrier, we shut up at each other. Shutting up *at* a person can be just as aggressive and deadly hostile as screaming at him. Indeed, people sometimes *threaten* to shut up! Yes, it sometimes happens that a person in company says: "My trouble is that I talk too much! I always get into trouble when I open my mouth. What I must learn to do is to shut up! Yes, to SHUT UP!! But don't worry; I'm learning!" And so everyone else in the company feels glum and guilty and worries that maybe *he* has been unjust and unkind to the "shut-upper" and that he is responsible for the shut-upper's drastic decision to shut up.

I know of a still more curious situation: There are two people—call them A and B. Person A dogmatically states a fact which Person B knows is false. And so B points out A's error. Then A descends on B with the fury of hell and says: "Why do you always contradict me? Why can't you let me say what *I* want to say? Why must you always shut me up?" Person B then remonstrates: "I have contradicted you, yes! But that does not mean that I am trying to shut you up. I objected only to the *content* of what you said, not to the fact that you *said* it! Of course you should be free to say what you want, but by the same token I should be just as free to say the opposite. Why, when I contradict you, do you interpret it as 'shutting you up'? If anything, it seems that it is really you who are trying to shut me up. I say this because you objected not to the content of what I said, but to the fact that I said it. You in effect told me I *shouldn't* contradict you. So who is really trying to shut up whom?" Person A then responds: "There you go shutting me up again! Now you are trying to make me feel guilty for having said what I said. Isn't this a way of shutting me up?" Person B responds: "I was not trying to make you feel guilty; I was merely pointing out to you what in fact you are doing." And so the argument continues, each party genuinely convinced that the other party is trying to shut him up. Yes, situations like this really exist, unfortunately.

There is much more I could say on the subject of "shutting up," but I think I have already said more than I should. I really should learn to shut up!

On Remaining Silent

☐ This essay is not self-contained. It presupposes my earlier essay "On Shutting Up." It can be appropriately regarded as a sequel.

Just what is the difference anyhow between shutting up and remaining silent? Do they mean the same thing? Well, they may have the same denotation, but hardly the same connotation! I cannot give too good a *logical* analysis of the difference, but I certainly have a *feel* for it! Consider, for example, the Chinese sages: It is said "The true sage remains silent." It would hardly do to say "The sage shuts up!" Also, a flower is silent. But it would sound most peculiar to say that a flower shuts up (except in a purely botanical sense). For that matter the Tao is silent (or at least it *says* it is; it is amazing how the Tao manages to *say* it is silent but says it so silently that it is true!). But to say the other about the Tao sounds so horrible to my ears that I cannot even bring myself to do so.

Now let us consider people. What is the difference between keeping silent and shutting up? It seems to me largely a difference of motivation and attitude. Perhaps I could define "shutting up" as "conspicuously remaining silent." I have already said that I regard shutting up—and even more so "threatening to shut up"—as something essentially aggressive and hostile. But "remaining silent" seems to me something more defensive than aggressive. Suppose you find that whenever you freely express your thoughts to another person—whenever you are truthful, honest, open, and "truly yourself"—the other person verbally hits you on the head—becomes intensely angry, furious, hostile, and vituperative. What do you do? Well, under these circumstances it might indeed be the wisest thing for you to tactfully remain silent, in the true manner of the sages! But this is very different from "shutting up"! If you are truly tactful, if you are truly a sage, then your silence will not be noticed; it will not *appear* that you are tactfully silent. Indeed if your opponent (?) feels that you are remaining silent on his account, in other words that you are shutting up, he will be more furious with you than ever! So you will have aggravated the situation rather than relieved it. I think this is a key difference between shutting up and remaining silent. The one intensifies the conflict; the other relieves it.

There are some other aspects of the situation to consider. "Shutting up" connotes a certain degree of violence both to oneself and others. "Shutting up," besides being a rather vulgar notion, has sadomasochistic overtones. Doesn't "remaining silent" sound somehow more serene and peaceful than "shutting up"? Also, "shutting up" sounds more *active*; "remaining silent" seems more passive. "Shutting up" also seems to involve a schism in the psyche; there is the "shut-upper" and the "shut-uppee." Shutting up involves "effort" and "discipline"; remaining silent does not. Remaining silent involves rather the Taoistic "wu-wei" effortless action or "action through inaction." When one shuts up, he "does something"; when one remains silent, he merely remains silent. In the words of one of the profoundly wise:

> The sage is quiet because he is not moved,
> Not because he wills to be quiet.

Praise or Blame?

☐ Ideally I would like to be praised for the good things I do but never blamed for the bad things I do. It would be nice if whenever I did something good, I could proudly praise myself and pat myself on the back and say "I did it!" but whenever I did something bad, I could take refuge in some form of "determinism" and say things like: "It was not really *I* who did it; it was 'fated,' it was determined by past causes; I had nothing to do with it, the action came not *from* me but *through* me." Yes, this is the attitude I would like to have. I realize, however, that this is hardly a fair one! A more realistic and fairer question would be "Would I rather be *both* praised for the good things I do *and* blamed for the bad things I do, or would I rather neither?" My answer is unhesitatingly and wholeheartedly "Neither."

In actual fact, though, I think I tend to praise myself for the good things I do but not to blame myself for the bad things I do. Or is it that I really do the reverse and only *wish* I did this?

On the Self-Perpetuating Nature of Anxieties

☐ I wish to give a few simple illustrations of a common theme—one method by which anxieties "protect themselves."

1. A girl once told me about her extreme fear of taking airplanes. Someone else present suggested that perhaps her fear could be overcome. At which I said: "But I bet you're *afraid* of losing your fear, aren't you? Because if you lost your fear, then maybe you actually would take a plane and get killed!"

2. I knew a man who was afraid of death. He was also afraid that if he lost his fear of death, he might fail to take those precautions necessary for him to survive.

3. I knew a Catholic who believed his salvation to be uncertain. He believed he had done everything necessary *so far* to ensure his salvation, but he had no guarantee that at some future time there might not come his way some evil temptations which he might not be able to resist. A Fundamentalist present, who was sure that he himself already *was* saved, said, "The Catholic Church really encourages anxiety, doesn't it?" My Catholic friend replied, "Of course, it's supposed to." He continued the position (which I think he got from C. S. Lewis)

that one has to be all one's life feeling on the edge of a razor between salvation and damnation, and that one was never safe nor could relax; salvation requires a *continually renewing attempt*. In other words, anxiety all one's life is the price one must pay for salvation. If one loses this anxiety and becomes "smugly assured of one's salvation," one might thereby lose it.

I am not accusing the Catholic Church of this point of view, I merely cite it as *his* version of Catholicism, and bring it up merely to illustrate another case of one's being anxious about losing one's anxiety.

4. Sometimes one has anxiety that someone else is untrustworthy, and he fears that if he loses the anxiety he will trust the other one and be betrayed.

The general formula is that one is afraid of something and that only by fearing it can one protect oneself against it.

Anxiety is really vicious! It is like a slave owner who says, "If you think you are badly off *now*, just try to rebel against me and you will see how worse off you will be!"

Closed Systems of Thought

☐ One of the human phenomena I find most disturbing is that of a person whose system of thought is such that there is no possibility of his ever finding out that he is wrong—even if he is. Any rational objection to his system can be explained away by a rationalization within the system, whose validity can be known only when one accepts the very premises of the system which are in question. Consider some obvious examples:

1. A Calvinist who, when questioned as to the fundamental tenets of Calvinism, will exclaim: "Of course you cannot see that I am right. Your trouble is that you haven't been saved!"

2. A Dogmatic Theist who, when questioned as to God's existence, will say: "Of course you cannot believe in God! You are too *proud* to admit the existence of a being greater than yourself."

3. A believer in the existence of the Devil who will say: "Of course you don't believe in the Devil. The first thing the Devil cleverly does is to convince people he doesn't exist."

4. An Atheist who will say: "No rational argument I can give you will convince you there is no God. You have a childish, superstitious need to believe in one."

5. A Marxist who will say: "Of course you cannot accept the Economic interpretation of History nor realize that the Class Struggle is *the* central issue. Your upbringing has been too bourgeois."

6. A Freudian who will say: "Of course you cannot see that I am right. All the reasons you have given against psychoanalytic theory are purely defensive rationalizations against realizing that which to you is most threatening."

7. A Feminine Liberationist who will say: "Of course you cannot realize that this is a man's world and that men are dominating women, not only on the economic level, but equally on the personal and psychological level. Of course you cannot see this; you are a man!" [Or, if she is speaking to a woman: "Of course you cannot see this; you have been dominated by male chauvinist ideology, which only proves my point!"]

I have perhaps given more than enough examples. The interesting thing is that in the majority of the cases, each of the groups I have mentioned can easily see through the prejudices of the others. And surely I must be in a similar category without realizing it. I wonder what *my* prejudices could be?

A Thought on Anxiety

☐ Yesterday I was having dinner with a group, and the discussion turned to the subject of anxiety. One member said that we can't get away from anxieties in life; anxieties are necessarily part of the very life process. Now usually when I hear such remarks, I get "up tight" and in a very superior—almost condescending—manner I hold forth and ride my favorite hobbyhorse, which is to the effect that anxieties are *not* necessary in life; life can and should be lived without anxiety. I still believe this as firmly as ever. I do not claim that I personally have yet found a way to live without anxieties; I merely believe that such a way *can* be found. I say this because I have consistently observed that whatever worthwhile things I have done in life I have done without any anxiety whatever, whereas whatever anxieties I have had have all been to no useful purpose. Thus I like to tell people that it is but a superstition to believe that anxieties are necessary. But last night I remained strangely silent through the conversation. One reason is that the following thought occurred to me.

Relief of anxiety (as well as all other sufferings) is my main interest in life. But I suddenly realized that for many people, the belief that anxieties are necessary in life, this very belief, is itself anxiety relieving! So is it not irrational for me to tell such people that anxieties are *not* necessary in life when my telling them this only arouses anxiety on their parts?

When Your Friend Throws Darts at You

☐ You have a friend—or a husband or wife or parent or child or brother or sister or cousin, or something—who claims to love you—or who sometimes maybe *does* love you—who is constantly throwing darts at you. What are you to do? You are wearing a chain of armor, but it has certain chinks, certain weaknesses. Your friend knows you very well and is extremely adroit in finding the weaknesses and inserting the darts where they really hurt. What are you to do? I can think of several possibilities. One is to let him keep hurting you and to bear the pain as gracefully as you can: indeed, don't even let him know that he is hurting you, because it might "hurt his feelings" to be told that he is hurting you. Well, this is the first possible course one can take. It is definitely *not* the course I recommend. Some may call such a course "saintly," but I think this is a totally perverted notion of saintliness. I may return to this point later.[1]

A second obvious course is to make your own darts and to throw them back at him every time he throws darts at you. This course is not too uncommon. But it is hardly without disadvantages. For one thing, it can lead to perpetual warfare. For those who enjoy warfare, this may be the best course. But I presume the majority of my readers care more for peace. It should also be remembered that your friend *may not even be consciously aware that he is throwing darts*! The capacity of the human mind to rationalize is fantastic! Not every sadist *knows* that he is a sadist! If your friend were *told* that he is throwing darts, he might become highly indignant. He might say: "Who, *me? I* never throw darts! I am a peace-loving man! Why should I throw darts? It is really *you* who are throwing the darts! I am only defending myself." And so now if you start throwing *real* darts at him, he may be overjoyed! He will say: "See, so I was right! Now you cannot *deny* that you are throwing darts! It was really you who was throwing the darts all the while. I was only defending myself, as I said!"

Let us now consider a third approach, which is my favorite one. This is to carefully look over your own armor, inspecting it

[1]In actual fact I don't!

for all possible chinks and weaknesses, and then to carefully reinforce every one of them. Thus fortified, your friend can throw at you all the darts he likes, and they will harmlessly fall from your armor like water from the back of an oiled duck. Then you will be *free*! Free at last! Free from the arrows of outrageous misfortunates (or whatever the saying is)! But, I am afraid, your friend will resent this most of all! You will *really* hurt his feelings! He will tell you that he preferred the "good old days" when you threw darts back at him. At least then you were a *man*; now you are only a coward and a cop-out. At least *then* you had some sort of *relationship*, even though the relationship may have had its "problems." But after all, are not "problems" part of a close "relationship"? How can you have a beautiful "close" relationship without sometimes getting angry (I would call it "spiteful," but your friend will call it "angry"), without getting "angry" and throwing darts? What's the matter with a few darts, anyhow, can't you take a "joke"? Why are you so "sensitive"? And now, instead of throwing darts back at me and trying to kill me in a spirit of good clean fun, you now "withdraw" from me; you "isolate" yourself from me, you "insulate" yourself, you "encapsulate yourself." Your trouble is that you have never had a really *close* relationship before! Now that you have had a "close" relationship with me, you are terrified that your real emotions will come out and you will have to *face* them. So instead you withdraw into yourself! But this can't work! Sooner or later you must face reality and realize that there are other people in the world too! No, you are too isolated, too encapsulated; you run away from every "encounter" that you face. What you are really doing is running away from life. But you can't keep doing this forever! You should see a psychiatrist!"

Yes, indeed, your friend will say all that! What can you do in such a world? I wish I could tell you. I wish I could help you. But I am unfortunately not yet a complete sage.

A Thought on Parents and Children

☐ Parents sometimes do things which the children absolutely hate, assuring them (as well as other adults) that these things are really *good* for them. The parents say, "We are older, wiser, and more experienced than our children and often know better what is good for the children than the children know themselves." And the parents say to the children: "You are too young *now* to realize that what we are doing is really in your best long-term interests. But when you grow up and become 'mature,' *then* you will realize that we were right."

It must be awfully embarrassing to the parents when the children grow up and continue to believe as firmly as ever that the parents were *not* right! Say the child reaches seventy, and the parents are somewhere in their nineties, and the "child" keeps saying: "Nope! All my life experiences have convinced me more than ever that you were totally wrong!" If the parents would say, "Well, we did the best we could," I'm sure the "child" would forgive them on the spot. But suppose the parents persisted: "We *were* right, only you were too young to realize it at the time." Now suppose the "child" replies: "But I thought you said that when I grew up, I would realize you were right. Well, now I have grown up. How come I don't realize it?" It is at *this* point that the parents must find the situation a bit on the embarrassing side! What can the parents possibly say? The only reply I can think of is this: "We did not say that! What we said was that when you grow up *and become mature*, then you will realize we were right! Now, in your case, you have grown up but clearly have not become mature. True, you are now seventy, but you still think like a child. The very fact that you still can't see we were right only proves how immature you still are!"

This argument, I admit, is unbeatable!

On Cutting One's Hair

☐ *Adult:* And therefore you should cut your hair. Maturity consists in the realization that we live in a society, and we can't all do what we like. It would be a fine how do you do if we all went around doing just what we pleased! There are other people also to consider. As you get older, you will realize that life can't always be the way you like. We must learn to *compromise.*

Teenager: Do you really believe me that simpleminded as not to realize that we can't all do what we like? Do you really think that lowly of me as to believe that I would really want to gratify all my desires—even those which hurt others? Of course I realize that I can't gratify my every whim, nor would I want to. But I do not regard this limitation as a "compromise." This limitation (if it should be called such) is something *I* want. Indeed, to call it a *compromise* strikes me as quite horrible!

Adult: Call it what you like. It *is* a compromise whether you call it a compromise or not. It makes no difference what *word* you use for it; the fact is that you have to compromise whether you like it or not.

Teenager: I have nothing more to say about this than what I have already said. However, the question of "cutting one's hair" does not come within the scope of what we have been discussing. If I am doing some act which *hurts* others, then I grant you have a legitimate right to complain. But how can my letting my hair grow possibly injure others?

Adult: Of course it hurts others! If you were living in a society in which everyone grew his hair, I would not say a word. But you are not. You are living in a society in which short hair for men is the accepted norm. And your refusal to abide by this norm is a direct act of aggression against society. And think of your parents and those who care for you! You don't

WHY IS LIFE SO RIDICULOUSLY PARADOXICAL? ————————

think it hurts *them* seeing you walk around like this? Could you not have the kindness to cut your hair at least during the years you are under our roof and providing for you, and later on in life, when you are on your own, then you can make your own decision.

Teenager: Okay, if my long hair hurts you *that* much, then I'll be happy to cut it.

Adult (In great alarm!): No, no! I don't want you to cut your hair to please *me!* I don't want to be the one to make you cut your hair. I want you to cut your hair because *you* want to, not because *I* want you to!

Teenager: Then I am afraid that you are asking for the logically impossible. If you ask me to cut my hair, then it is *possible* for me to do so. But if you demand that I *want* to cut my hair—that I would cut my hair even had you not asked me—that is simply impossible.

Adult: Then I think you should see a psychiatrist. I'm sure that if you were in the hands of a competent analyst, then after a couple of years you would be sufficiently mature to *want* to cut your hair. You would then do so of your own accord without anyone's telling you to.

Teenager: But I don't *want* to want to cut my hair!

Adult: You are playing silly word games! If you went to a psychiatrist, you would, after a while, want to cut your hair, and you would be glad that you wanted to.

Teenager: But since I don't want to want to cut my hair, then if, as you say, a psychiatrist would influence me in the direction of wanting to cut my hair (which, incidentally, I don't believe would really happen, but let us assume for the moment that it would) then, as I say, since I don't want to cut my hair, and a psychiatrist would influence a change in this "want," then I obviously don't want to see a psychiatrist.

Adult: But you *should* see a psychiatrist.

Teenager: But I don't want to.

Adult: There you go being selfish again! Because *you* don't want to see a psychiatrist, you don't see one. It never occurs to you that with your immaturity and emotional instability you are hurting countless others. But no! Because *you* are satisfied with your condition it follows that you don't change it.

Teenager: All right, if I am really hurting you with my so-called "neurosis," and you really want me to see a psychiatrist, I will see one.

Adult: No, no! I don't want you to see a psychiatrist just to please *me!* I want you to see a psychiatrist because *you* want to, not because *I* want you to.

Teenager: I'm afraid we are in the same position as before. The only way I can see out of the difficulty is this: Do you know of any other kind of being whom if I saw first, I would then *want* to see a psychiatrist, and also whom I would now want to see?

Adult: More word games! I wish you would realize this is a serious matter. Your jokes are not funny!

Teenager: I am being quite serious. I seriously cannot make out what it is that you *do* want me to do. You say you want me to cut my hair but only if *I* want to, which you know I don't. And you want me to see a psychiatrist, but only if I want to, which again I don't. So I am quite confused! What *do* you want me to do?

Adult: What do *I* want you to do? *I* don't want anything! I want you to do what *you* want to do, not what *I* want you to do.

Teenager: But that is exactly what I am doing. So what's the problem?

Adult: What's the problem? The problem is that you are not cutting your hair!

Teenager: But you told me that you don't want me to cut my hair unless I want to, and I told you I don't want to, and hence you don't want me to, so I am not doing anything you don't want me to do.

Adult: Aren't you clever, with your logic and sophistry! Look, I'm not here to discuss Aristotle and Plato with you.

Instead of spending all this valuable talent and energy on these senseless word games, wouldn't it be simpler for you just to cut your hair and look like a decent human being for a change?

Teenager: But I don't *want* to cut my hair!

Adult: But I *want* you to want to cut your hair!

Teenager: But I *can't* want to cut my hair!

Adult: What do you mean you *can't*? You mean you *won't.* Of course you could cut your hair if you wanted to!

Teenager: I didn't say I can't *cut* my hair. Of course I can! I said I can't *want* to cut my hair.

Adult: Then you should see a psychiatrist.

Teenager: Must we go through that again?

Adult: No, we don't have to go through *anything* again. I am perfectly satisfied with things as they are.

Teenager: Good. Can I go now?

Adult: Of course you can go now! Why do you have to ask my permission? Are you a child? You act as if I am restraining you. You can go whenever you like. Only I wish you would cut your hair.

Teenager: Tell me honestly, why is it so important for you that I cut my hair?

Adult: It's not important for *me*, it's important for *you*! As for *me*, I don't care what you do. You are no longer a child. You are old enough to make your own decisions. And you should decide to cut your hair. It is *you* who should decide it, not *me* who decides it for you.

Teenager: But *why* should I decide to cut my hair?

Adult: I shouldn't have to tell you why. You should know why by yourself, without my having to tell you.

Teenager: I'm sorry but I don't know why.

Adult: Of course you know why! How can you go around like that when you know it offends people?

Teenager: I'm sorry if it offends people. But I believe that I have the right to decide how I shall wear my hair.

Adult: You're *not* sorry it offends people! You are only too happy to offend people, otherwise you would not go around looking like that.

Teenager: No, that is not true. I do not do it *because* it offends people. I do it despite the unfortunate fact that it offends people.

Adult: But do you think it is right to go around offending people?

Teenager: There is no yes or no answer to that question. It obviously depends on the nature of the offense.

Adult: Well, would you walk around naked on the streets if you felt like it?

Teenager: Of course not.

Adult: Why not?

Teenager: In the first place, I cannot imagine that I would feel like it.

Adult: Why not? What's so shameful about nudity?

Teenager: There is nothing objectively shameful about it. It's just that my culture has induced in me a certain natural modesty which would make it impossible for me to feel like it.

Adult: But suppose you *did* feel like it. Would you?

Teenager: As I said, it is very hard for me to imagine that I could feel like it. But if I did feel like it? No, I'm sure I wouldn't anyhow!

Adult: Why not?

Teenager: Because I would not want to offend others.

Adult: Oh boy! Speak about inconsistency! Why is it any worse to offend others with nudity than with long hair?

Teenager: I don't know why. It just seems worse—much worse—to me.

Adult: Then you are being totally irrational! You disappoint me. I thought at least that you were intelligent.

Teenager: I'm sorry, but I cannot agree that I am being irrational. Just because I prefer one thing to another and am unable to state any *reason* for my preference does not mean that the preference itself is irrational. Nor am I irrational or inconsistent just because I don't know what the reason is. Does this mean that I am therefore unintelligent? Possibly yes. Perhaps if I were mentally quicker and could think better on the spur of the moment, then possibly I could come up with a fruitful analysis of why I prefer one to the other. But I admit that I cannot—at least at the moment—do this.

Adult: Your whole trouble is that you won't conform! You won't accept the mature discipline of conforming with the not unreasonable wishes of others. Again it is your ego. It is always "I," "I," "I." Because *you* like long hair, it makes no difference how others feel.

Teenager: I don't believe conformity has anything to do with it.

Adult: Of course it does! But what you don't realize is that although you are trying not to conform, you are actually conforming in the very act of letting your hair grow.

Teenager: What on earth are you talking about?

Adult: I mean that in letting your hair grow, you are conforming not to adult society, but to the society of your own age group. You grow your hair long *to be one of them.* So it's not a question of conforming or not conforming. You *have* to conform whether you want to or not. The realistic question is not *whether* you conform, but to which group you conform.

Teenager: At this point I must tell you that it is *you*, not I, who is playing word games. In particular, you are obviously using the term "conformity" in a nonstandard sense. True enough, I do like my own age group more than I like adults. And true enough, they like long hair. But since I myself like long hair, I am not wearing it out of conformity with my peers, but merely because I like it that way. If I actually preferred short hair to long hair, then I would indeed be conforming to my peers in wearing long hair. Or if I preferred long hair but wore it short just for the sake of the so-called "adults," then again I would be conforming. But since I am wearing it the way I like, then I am not conforming. To "conform" means to go *against* one's natural inclinations for the sake of another's approval. Therefore you are totally confusing the issue when you use the term "conform" in the way you did.

Adult: Again you play all this semantics on me! I told you I'm not here to discuss with you the philosophy of words. You should get your hair cut! Don't you realize that your long hair makes you look *effeminate!*

Teenager: Oh, that's what's bothering you! Why didn't you say so earlier? Well, I don't believe for one moment that it makes me look effeminate. To me it is inconceivable that anything which naturally grows on a male body can make it look effeminate. Next, perhaps you will be telling me that my sexual organs look effeminate. Perhaps you would like me to remove them also?

Adult: Now stop being facetious!

Teenager: I'm not being facetious. I quite seriously would not be surprised if society is moving to a point of perversity in which men will get the "brilliant" idea that their sexual organs look effeminate and hence will remove them in order to look more "masculine."

Adult: Enough of these sick fantasies! You know they are ridiculous! I am not trying to "castrate" you, but just the reverse. To me, a man should be a man, and a woman should be a woman.

Teenager: What a sad state of affairs that people should have to talk like this! To me, a man *is* a man, and a woman *is* a woman, without having to believe they "should" be. I think all this overanxiety about men being masculine and women being feminine is doing far more harm than good.

Adult: Look, I think it is pointless to argue any more with you about this. You stubbornly refuse to cut your hair, and obviously nothing I can say will change your mind. I think even if God himself came down to earth and told you to cut your hair, you still wouldn't.

Teenager: Of course I wouldn't!

Adult: So even God himself can't influence you!

Teenager: That does not follow. That does not follow at all!

Look, in the first place, I don't know whether or not there is a God. But let us for the sake of argument assume there is. Then, as you say, if he came down to earth (which, by the way I think is a ridiculous concept, but let that point go)—if he came down to earth and commanded me to cut my hair, I would refuse. This means only that I am not sensitive to what might be called the "external" voice of God. However this does not imply that I am insensitive to the "internal" voice of God—to the voice or influence which works *within* me. If God really wanted me to cut my hair, he would know better than to come down to earth and "tell" me to do so; he knows me well enough to know that under those circumstances I wouldn't. If he wants me to cut my hair, then he will cause me to *desire* to do so. It is that simple.

Adult: You speak as if you have no will of your own! There *is* such a thing as free will, which God has given you. It is *you*, not God, who decides whether you cut your hair.

Teenager: You raise a subtle but extremely important point. You are partly right and partly wrong. Yes, it is indeed *I* who decide whether I cut my hair. But that does not at all rule out that God also decides whether I cut my hair.

Adult: I don't understand you at all!

Teenager: Let me put it this way. If I am hungry, then it is up to me whether I eat or not—*I* then decide whether or not to eat. But I have never in my life *decided* to be hungry. Similarly, if I have the *desire* to cut my hair, then it is up to me whether I cut my hair or not. Or if I desire *not* to cut my hair, it is again up to me what I do about it. But it is *not my* efforts, *my* will, or *my* power which determines my very desires. Whatever this power is—call it Nature, God, or what you will—it is *this* power which is the primary element in deciding whether or not I cut my hair. People do not recognize the importance of this power; they think it is "they" who do everything. This is what is sometimes referred to as the sin of "pride" or "arrogance." Really, it is more the result of ignorance.

And your sin, my dear parent, if I may be so bold as to tell you, is that you have constantly tried to usurp this power. You have tried to take the role of God in my life.

Adult: What on earth are you saying?

Teenager: What I say is absolutely true, though I realize it will be extremely difficult and painful for you to accept. Yes, you have indeed attempted to usurp God's role in my life, and this has been the major cause of all our psychological difficulties. Recall earlier parts of our conversation. It was not enough that I finally agreed to cut my hair to please *you*; you were totally enraged and frustrated that I would not cut my hair to please *myself.* In other words, it was not enough for you to bring various pressures on me to change my *behavior*; you tried to force a change in my very *desires.* To try to make me cut my hair is not usurping God's role—it is quite normal (unfortunately) for parents to do *that.* But to be infuriated with me, not for not *cutting* my hair, but for not *wanting* to cut my hair, is something far more drastic and far more damaging. It is really commanding something impossible. It makes sense for you to command me to cut my hair. But it is senseless—far worse than senseless—to command me to *want* to cut my hair. It is this latter type of command, not the former, which is a usurpation of what might be called "God's role."

Adult: Now don't be ridiculous! How could I possibly command you to *want* to cut your hair? I can't get inside you and make you "want" what I want you to want. Who do you think I am, God?

Teenager: No, that is the whole point. But you frequently act as if you were, or rather that you *wish* you were. I have long felt that your main problem in life is that you are not God, you know that you are not, and you are furiously raging that you can't be. (That's perhaps the *real* reason you left the church!) I say this not only on the basis of your behavior toward me and the other children, but also your behavior to your spouse and many of your friends. I've constantly noticed that it is not enough for you to bend people's *actions* to your will, but you rage and fume and act hurt that you can't bend people's very "will" to your will. And this, I say, is your major problem.

I'm sorry that I cannot be of more help to you in this matter. I hardly expect you to believe what I'm saying as it is. And now I really *do* wish to take leave.

III I TOLD YOU SO!

A Strange Paradox

☐ I think this is implicit in much of our thinking. Consider the man who says, "Oh God, I'm nothing! I'm nothing! I'm only an insignificant speck in a vast universe. I'm really nothing!" But this same man will also say, "I am human, and humans are of course superior to animals (for God has privileged us by giving us free will), and animals are of course on a higher level than plants. Now, flowers are plants. But a flower! What could be more beautiful and perfect than a flower? A flower is not lacking in *anything*! A flower is as beautiful and perfect as it can be. It is God's creation at its best. It cannot be any better than it is."

So here am I obviously better than a flower. A flower is perfect, yet I am only a miserable nothing. Is this not remarkable?

Four American Indians

☐ *First Indian:* Do I *believe* in the Great White Spirit? Of course I believe in the Great White Spirit. I even *feel* the Great White Spirit.

Second Indian: How I envy you! I have always *believed* in the Great White Spirit but never could *feel* the Great White Spirit.

Third Indian: How strange! My case is the opposite. I have always felt the Great White Spirit but never could believe in the Great White Spirit. I believe that my *feelings* are but childish superstitions reinforced by my ancestral teachings.

Fourth Indian: It seems I am the only wholly mature and sane one among you! I neither feel the Great White Spirit nor believe in the Great White Spirit.

Do Animals Perceive God?

☐ Assuming there is a God, I would not be surprised if animals perceive Him directly whereas humans have somehow lost the ability to do so. In that case it may be that the *real* reason why many humans believe in a God is that they actually have dim archaic racial memories of the days in which they perceived Him directly.

I am not altogether kidding! I often have the feeling that my dogs Peekaboo and Peekatoo are in *direct* contact with the Divine which we humans merely "reason" about. If my conjecture should be correct, then would not all the theological disputes of the past amount to one grand joke? One can reason against reasons. It makes sense to reason against faith. It *certainly* makes sense to reason against authority! But to reason against actual *memories*—even unconscious ones? What power could such reasons have?

A New Religion

☐ I believe religions have been invented by *men*. I doubt that any of them (even atheism) is wholly *true*. Yet each may contain some kernel of truth. Since other men have been allowed to present their speculations, why should I not be allowed to express mine? After all, is any one person really better than any other? So here is my speculation. [I have not adequately checked the literature, so I do not know how original it is.]

There is a God. But God is *not* immortal. His life span, though, is very long—say some millions or billions of years. After God goes, He will need a replacement (to keep the universe going). So who will the next replacement be? *Answer:* us! The whole purpose of evolution is to train us mortals (?) to be the next God. This is our *real* purpose! After countless millennia, in which, say, we transmigrate over and over again, we develop to the point where we all fuse together and become the next God. Then we take over the divine functions, start training mortals to replace *us*, etc., *ad infinitum*.

Is Life Tragic?

☐ I understand from many highly learned people that life is essentially a tragedy. Not only is it tragic in virtue of the enormous sufferings of life, but then comes the Grand Tragedy that we must all die in the end! We strive to live as long as we can, but alas, we fail to live eternally, and so Death is our final defeat! My learned friends also inform me that not only is it *they* who have reached this profound wisdom, but if only I would *read a little more* I would find out that *all* the great and sensitive thinkers of the past have sooner or later come to this realization. I am also informed that *my* inability to realize that life is tragic is due not only to the fact that I have not read enough, but also to the fact that I have not *lived* enough, and to my general immaturity, coupled with the fact that I look at life through rose-colored glasses, that I stubbornly refuse to see that death *is* a tragedy, and also that my own life has been unusually fortunate and that I haven't personally suffered enough.

Well, who am I to contradict the word of so many great intellects? But I must confess that I have never in my life thought of life as tragic! Even in my most depressed and suicidal moods (which I imagine hit almost everyone at times), my attitudes toward this matter do not change. When I have had suicidal thoughts, it is not that I thought that *life* is tragic, nor even that *my* life is tragic, but simply that I am in a temporary state of excruciating pain and I want the pain to end. In other words, it's not that I want to be dead *per se*, it's just that I don't want to be hurt. But as to life's being *tragic*, the very thought that life is tragic strikes me of all things as extremely *funny!*

From the little I *have* read, and from what I have observed of people firsthand, I strongly suspect that there is very little correlation between how much one has suffered and how strongly one feels that life is tragic. I think the "tragic sense of life" is rather a temperamental affair which has little to do with the amount of suffering one has either experienced or witnessed. I think it is the *type* of suffering rather than the *quantity* of suffering which is relevant. To put the matter quite bluntly—and perhaps a little too harshly—I cannot help but suspect that those who harp most on the tragedy of life are the very ones who introduce most of the tragic elements that are in life. That is to say, there are those whose attitudes are such that they are bound to be unhappy regardless of how favorable is their external environment. And, I think, it is *these* people who talk most of the tragedy of life, not those who actually suffer most.

The tragedies of life, so I am informed, are suffering and death. That death is a tragedy is something I doubt I will ever believe. Suffering might come somewhat closer (especially when the suffering appears to be preventable), but I think this still misses the mark. There is nothing I dislike more than suffering, but I feel that suffering is something to be *prevented, helped, healed, eliminated*, rather than something to be "trag-edized" over. Suffering is just a God damn pain in the neck!

I think the word "tragic" as used by those who claim life to be tragic connotes to me a certain romantic and melodramatic attitude with definite sadomasochistic overtones. I doubt that the Greek tragedians ever though of life as "tragic" in this way! Their "tragedies" are of a totally different spirit. The beauty of them does not arise so much out of the *suffering* involved, but rather out of an essential feeling for Universal Harmony. Christianity seems more to utilize suffering as an essential ingredient. Most of all I admire the beauty of much Far Eastern thought, whose sublimity arises not *through* suffering, nor in indifference to suffering, but quite *independently* of suffering. (In this way, it greatly resembles the sublime beauty of Science and Mathematics.)

So in response to those who claim that the lack of "tragic sense of life" is indicative of immaturity and lack of emotional depth, I wish to go on record as making the counterclaim that the highest form of beauty is totally unconcerned with the negative aspects of life.

It may strike some of my readers as inconsistent with everything I have been saying when I state that I love to read the so-called "pessimistic" philosophers such as Schopenhauer and von Hartmann. They are really among my favorites! I once asked a literary philosopher why it is that I find the pessimistic philosophers not depressing or anxiety producing, but infinitely *soothing*. He gave the interesting answer, "I think this is because the pessimistic philosophers are really optimists at heart." I also once asked another philosopher, "Why is it when I read the pessimistic philosophers, I feel so cheered up?" He said, "Of course, because you know it isn't true."

A Thought on Schopenhauer

☐ That Schopenhauer should be regarded as a *pessimist* strikes me as fantastic! Maybe he is somewhat pessimistic about Life (though not in a very convincing way) but he sure as hell is incredibly optimistic about Death! Where in the world's literature do you find anything more optimistic than the following:

If now the all-mother sends forth her children without protection to a thousand threatening dangers, this can be only because she knows that if they fall, they fall back into her womb, where they are safe; therefore their fall is a mere jest.

Some readers will claim: "That *surely* is optimistic! But totally unrealistic. Schopenhauer is acting as if he believed that after death we still in some sense *exist* and are 'safe in the womb of Mother Nature.'" I myself would give a more radical interpretation. I think that the question is not whether we exist after death, but whether existence itself is all that different from nonexistence. Some irate reader will reply: "Of course they are different! Existence is the very opposite of nonexistence!"

I know they are opposite! But so is minus zero the opposite of plus zero.

I Told You So!

☐ The following conversation occurred while I was driving with some close friends including one man—call him "B" — and a girl—call her "M." I had recently been discussing with B the question of an afterlife. He believed it to be a sheer superstition. On the other hand, M tends to believe in some form of afterlife on the grounds—shared by Goethe—that a person's own nonexistence is inconceivable to one, and that it is impossible for one to believe in something he cannot even conceive. I tend to agree with M, and I furthermore believe that the belief in the reality of death is but a totally unfounded superstition. I had also been recently discussing with B the question of ego-assertiveness. B thinks that I am very ego-assertive as proved, e.g., by my attitude toward ego-assertiveness. More specifically, he thinks that in my very condemnation of ego-assertive people, I am only asserting my own ego. I think that *he* is being ego-assertive in asserting this, but enough! At any rate, all these thoughts were on my mind at the time of the conversation. The conversation went as follows:

I opened by saying: "Would you like to know the *real* reason I hope there is an afterlife?"

B replied: "Yes?"

I said: "So I can then triumphantly say to all my skeptical friends, 'I told you so!'" This got a general laugh.

Then B said: "But wouldn't it be funny if you were the *only* one to survive? Then you would have no one around to gloat over!" This, of course, got another laugh.

Then M said: "It would be interesting if just those who believed in survival survived, and those who didn't, didn't."

I thought about this and fantasized myself surviving with all the other believers and being totally frustrated at not being able to gloat over any nonbelievers, since they would all have, so to speak, "gone out of existence." So I would be foiled again! Finally I said: "I think it would be still funnier if the reverse were true—if the skeptics were the only ones to survive. I can then imagine all the skeptics together saying: 'Poor Raymond! Too bad he is not with us, gloating and telling us, "I told you so." He sure did tell us so! If he hadn't, he would be here with us now!'"

IV PHILOSOPHICAL FABLES

The Astronomer Who Believed in the Possibility of Miracles

☐ Once upon a time there was an astronomer. He believed that the laws of celestial mechanics *usually* held. But he was open to the possibility of miracles. He did not go so far as to believe that there ever *had* been any miracles, or that there necessarily *would* be any miracles in the future; he was merely open to the logical possibility that there *might* have been miracles in the past, or that there *might* be miracles in the future. This astronomer knew his celestial mechanics very well, but his temperament was of the skeptical type; he was a sort of "doubting Thomas." Of course his doubting Thomasness was not of the usual type, for whereas the Biblical doubting Thomas was doubtful about miracles, *he* was doubtful that it really was known that miracles never occur. He said: "Just because I never saw a miracle, or don't know of any miracle that actually happened, why should I believe that miracles never happen? What *proof* do I have that there are no miracles?" And so whenever he made any astronomical prediction, he usually prefaced it or appended it with "unless there is a miracle."

One day he predicted an eclipse. People asked him, "Are you sure it will take place?" He replied, "Of course it will, unless, of course, there is a miracle."

The day of the predicted eclipse arrived, but there was no eclipse. The people excitedly went to him and asked: "What happened? Was there a miracle?" He replied: "No, no; there was no miracle. It just so happened, I miscalculated."

The Seer and the Skeptic

☐ *Seer:* In yonder bush there is a monster with three ears, five legs, two tails, and orange eyes!

Skeptic: Nonsense! In that bush there is no monster with three ears, five legs, two tails, and orange eyes!

Seer: There is indeed such a monster in yonder bush.

Skeptic: There is no such monster in yonder bush.

Seer: You don't have to take my word for it, all you have to do is go over and *look* into the bush, and you will see such a monster.

Skeptic: I will be happy to take a look for myself, but I can assure you I will *not* see any such monster.

Seer: It is futile for us to discuss this further until you have taken a look and seen for yourself.

Skeptic: Very well, I will go now and look.

A few minutes later

Skeptic: Ha! ha! I was right! There is no such monster in the bush.

Seer: Really? That is fantastic! I could have *sworn* I saw such a monster in the bush.

Skeptic: That is the trouble with you seers! You are always *sure* you see things which in reality don't exist.

Seer: I still can't believe there is no such monster in the bush!

Skeptic: I can assure you there isn't!

Seer: Did you look carefully?

Skeptic: Most carefully.

Seer: And you saw nothing?

Skeptic: Oh no, I saw something.

Seer: What did you see?

Skeptic: I saw a monster with three ears, five legs, and two tails. But its eyes are yellow, not orange!

A Sad Story

☐ Once upon a time there was a man who spent the first half

of his life trying to become famous. He failed in that. Then he spent the second half of his life trying to get into that mystical state in which it was no longer important to him whether or not he was famous. He failed in that too.

Another Sad Story

☐ Once there was a man who was overcome by mystical inspiration. He had all sorts of remarkable insights as to the ultimate nature of reality. He wrote voluminously; he wrote and wrote and wrote. He was not, however, entirely egoless, for he took great pride in what he had written. For many months after he had finished writing, he read his manuscript over and over again with great pride and joy.

During the next couple of years he slowly but surely lost all of his mystical insight. Then one day, he reread his manuscript but could not understand a word of what he had written.

An Unfortunate Dualist

☐ Once upon a time there was a dualist. He believed that mind and matter are separate substances. Just how they interacted he did not pretend to know—this was one of the "mysteries" of life. But he was sure they were quite separate substances.

This dualist, unfortunately, led an unbearably painful life—not because of his philosophical beliefs, but for quite different reasons. And he had excellent empirical evidence that no respite was in sight for the rest of his life. He longed for nothing more than to die. But he was deterred from suicide by such reasons as: (1) he did not want to hurt other people by his death; (2) he was afraid suicide might be morally wrong; (3) he was afraid there *might* be an afterlife, and he did not want to risk the possibility of eternal punishment. So our poor dualist was quite desperate.

Then came the discovery of *the* miracle drug! Its effect on the taker was to annihilate the soul or mind entirely but to leave the body functioning *exactly* as before. Absolutely no observable change came over the taker; the body continued to act just as if it still had a soul. Not the closest friend or observer could possibly know that the taker had taken the drug, unless the taker informed him.

Do you believe that such a drug is impossible in principle? Assuming you believe it possible, would you take it? Would you regard it as immoral? Is it tantamount to suicide? Is there anything in Scriptures forbidding the use of such a drug? Surely, the *body* of the taker can still fulfill all its responsibilities on earth. Another question: Suppose your spouse took such a drug, and you knew it. You would know that she (or he) no longer had a soul but acted just as if she did have one. Would you love your mate any less?

To return to the story, our dualist was, of course, delighted! Now he could annihilate himself (his *soul*, that is) in a way not subject to any of the foregoing objections. And so, for the first time in years, he went to bed with a light heart, saying: "Tomorrow morning I will go down to the drugstore and get the drug. My days of suffering are over at last!" With these thoughts, he fell peacefully asleep.

Now at this point a curious thing happened. A friend of the dualist who knew about this drug, and who knew of the sufferings of the dualist, decided to put him out of his misery. So in the middle of the night, while the dualist was fast asleep, the friend quietly stole into the house and injected the drug into his veins. The next morning the body of the dualist awoke—without any soul indeed—and the first thing it did was to go to the drugstore to get the drug. He took it home and, before taking it, said, "Now I shall be released." So he took it and then waited the time interval in which it was supposed to work. At the end of the interval he angrily exclaimed: "Damn it, this stuff hasn't helped at all! I still obviously have a soul and am suffering as much as ever!"

Doesn't all this suggest that perhaps there might be something just a *little* wrong with dualism?

The Man Who Was Satisfied with Everything

☐ Once there was a man who was satisfied with *everything*! His philosophy of life was that it is best to be satisfied with everything. And the amazing thing is that he actually lived up to his philosophy. His attitude got many people extremely exasperated. But he was satisfied with that too. When asked, "But are you *never* dissatisfied with anything?" he replied: "Oh, I am often most dissatisfied with many things, but my dissatisfaction does not bother me at all. I am perfectly satisfied to be dissatisfied." "But are you not being inconsistent?" "Of course I am! I am satisfied with that too!" "Oh come now, are you really satisfied with *everything*?" "Why certainly!" "Are you satisfied with the fact that there was an Adolf Hitler?" "Of course not!" "Are you satisfied that there was a war in Vietnam?" "Of course not." "Are you satisfied with air pollution?" "Of course not." "Are you satisfied with the fact that the human race is threatened with annihilation?" "Of course not." "Then what on earth *are* you satisfied with?" "I am not satisfied with anything in particular; I am only satisfied with things in general."

Compromise

☐ Once upon a time two boys found a cake. One of them said: "Splendid! I will eat the cake." The other one said: "No, that is not fair! We found the cake together, and we should share and share alike; half for you and half for me." The first boy said, "No, I should have the whole cake!" The second said, "No, we should share and share alike; half for you and half for me." The first said, "No, I want the whole cake." The second said, "No, let us share it half and half." Along came an adult who said: "Gentlemen, you shouldn't fight about this; you should *compromise*. Give him three-quarters of the cake."

Bits and Pieces

The Long-Term View

☐ I believe it is a mistake to judge one part of a person's life or behavior apart from the total picture.

Suppose, for example, a man beats his wife. He suddenly takes up burglary, and lo and behold, he no longer beats his wife. Under these circumstances, I would say the burglary was a blessing, wouldn't you?

Judge and Defendant

☐ *Judge (to Defendant):* And therefore the court sentences you to be hanged by the neck until you are dead.

Defendant: That's all I need!

A Hedonistic Sadist

☐ *Friend:* I don't understand; why do you want other people to be in pain?

Hedonistic Sadist: Oh no, you misunderstood! It's not other people's pain I desire, but only the pleasure their pain gives me.

Crazy?
☐ *A:* Why do you call Jim crazy; does he actually have hallucinations?

B: No, that's his whole trouble—he only imagines he does.

A Skeptical Mystic
☐ *John:* Have you ever had any mystical experiences?

Jim: Oh, I have them all the time, but I don't believe a one of 'em!

A Paradoxical Rationalist
☐ Once there was a man who was constantly and irritatingly rational. When asked, "Why are you so rational?" he replied: "Because it is irrational to be so rational. Basically I am irrational—I love irrationality; the more the better. The most irrational thing I can do is to be as rational as I am. That is the reason I am so rational."

In Harmony with the Tao
☐ The above is reminiscent of the story of one who asked a Zen master, "How do you get in harmony with the Tao?" The master replied, "I am already out of harmony with the Tao."

A Buddhistic Bit
☐ Two friends were with me on the porch. One said to me, "Are you sufficiently Buddhistic to object to killing a bug?" I replied *no.* The second friend said, "But the bug is!"

On Life and Death
☐ *A:* Would you call Henry a life-affirming person?

B: Oh, definitely! He loves living. He is typically the sort of person who will go on living the rest of his life.

A: But surely he will die *sometime*!

B: But that doesn't seem to bother him! As he says, "Why should I worry about dying? It's not going to happen in *my* lifetime!"

On Isolation

☐ *A:* How come you never see people? Doesn't that make you feel isolated?

B: Oh, not at all! I am, indeed, isolated from people. But when I'm alone, I don't *feel* isolated. When I'm with people, then I'm painfully aware that there are people I am isolated *from.* And so then I feel isolated. But when I'm alone, I totally forget that there *are* any other people, and so I then feel no isolation whatever.

Famous?

☐ I was once discussing the subject of fame with a very astute friend. At one point he said: "It all depends on what you mean by *famous.* For example, would you call God famous?"

Revelations

☐ Why should I believe other people's revelations; I have enough trouble believing my own!

Another Buddhistic Bit

☐ *Friend (to Buddhist):* This morning your ears should have been burning! All of us were discussing what a great guy you are.

Buddhist: (Smiles without saying a word.)

Friend: Why do you smile in this superior fashion? Isn't it rather un-Buddhistic to relish flattery in such an obvious way?

Buddhist: That's not what I was laughing at. What struck me as so funny was the fact that the "I" which you seem so greatly to admire doesn't really exist.

How Sad!

☐ I once wrote the following verse entitled "How Sad!"

How sad the trees
How sad the flowers
They cannot see their own beauty.

Two years later, to my utter amazement, I came upon the following passage in Spengler's "Peoples, Races, and Tongues":

"It is a sight of deep pathos to see how the spring flowers craving to fertilize and be fertilized, cannot for all their bright splendor attract one another, or even see one another, but must have recourse to animals, for whom alone those scents exist."

I Think, Therefore I Am?

☐ *I think, therefore I am?*
Could be!
Or is it really someone else
Who only thinks he's me?

Determinism or Free Will?

☐ *Scene 1*

Existentialist: But you *shouldn't* make such a promise. You have no way of knowing that you will keep it in the future.

Victim: What the hell are you talking about? Of course I will keep it in the future!

Existentialist: But how do you know you will keep it in the future?

Victim: Because I have *decided* to keep it in the future. That's how I know!

Existentialist: I do not deny that you now honestly intend to keep your promise. It is just that I do not see how your present good intention can possibly guarantee that you will have equally good intentions in the future. After all, there *is* such a thing as changing one's mind!

Victim: Not with me there isn't! I never break promises!

Existentialist: How do you know that?

Victim: I have never broken a promise in my whole life!

Existentialist: I believe you. I don't doubt *that* for a moment. If you tell me that you have never yet broken a promise, I take your word implicitly. But just because you have never in the past broken a promise, why does it follow that you never will in the future?

Victim: I really don't understand you at all! Are you taking a completely deterministic or fatalistic position? Is it as if it were already "written in the book" how I will act in the future, and since I don't know the "book," I have no way of knowing what the future will be? In other words, are you trying to tell me that *I* have nothing to say in the matter?

Existentialist: No, no; I am not saying that at all. I happen to believe in free will. But strangely enough, free will is not the crucial issue here. That is to say, a strict determinist would come to the same conclusion I would, only for very different reasons. The determinist would say that whether or not you keep your promise in the future is strictly determined by the laws of physics together with the present configuration of the universe, and since you do not know enough about the present configuration, then you have no rational means of knowing the future. The determinist would thus say, "It is *not* up to you whether you keep your promise or not; it is up to the universe and its laws." I, on the other hand, believe in free will and hence believe that at any time it *is* up to you what you do at that time. You have now made a promise of your own free will. You presently intend to keep it. But I do not believe that you have the power to choose *now* what you will choose to do in the future.

Victim: But I *do* have the power! I told you I have never before broken my word. Does this not constitute very strong probabilistic evidence that I will not break my promise in the future?

Existentialist: Oh, probabilistic evidence! I wasn't thinking of this at all! I thought you were claiming it a *certainty* that you will keep your promise. If all you are talking about is *probability*, then I'm not sure I would disagree with you. Indeed, the fact that you have never yet broken a promise is in itself a good probabilistic indication that you won't in the future, though perhaps other relevant factors should be taken into account. At any rate, if all you are saying is that it is *likely* that you will keep your promise, I will not dispute you. All I object to is your absolute *certainty* that you will keep your promise.

Victim: Of course I am *certain*! It's not a mere matter of probability; it's a question of complete and total certainty. You forget that it is *I* who have made the promise. There is no power in the entire universe which can force me to break my promise.

Existentialist: I know that no power in the universe can cause you to break your word. I know that full well. You are one of the most strong-willed, self-willed people I have ever known—indeed, many would characterize you as downright "stubborn." And therefore I know that no external force can compel you to break a promise. But that does not mean that *you yourself* don't have the power to break it! And no one—not even you—can possibly know whether you will choose to do this in the future.

Victim: But I *won't* choose to do this! I *know* I won't.

Existentialist: But *how* do you know?

Victim: I don't know how I know, I just know!

Existentialist: Really now, that remark is hardly worthy of you! How can you say that you know but that you don't know how you know?

Victim: It is true. I *do* know, but I don't know how I know.

Existentialist: That is ridiculous! You *don't* know.

Victim: I *do* know!

Existentialist: You don't!

Victim: I do!

Existentialist: You don't!
   Etc., etc.

Scene 2

Several months later the victim fell into the hands of an evil brain surgeon. This surgeon was a diabolical character whose specialty was to perform brain operations whose express purpose was to make people choose to break their promises. And so the evil brain surgeon operated on our victim, and as a result the victim decided (?) to break his promise. Shortly after, he had the following conversation with a determinist.

Determinist: So! You broke your promise!

Victim: Oh yes!

Determinist: What would you say *now* to the existentialist? I know about your entire conversation!

Victim: Obviously I must admit that the existentialist was right and I was wrong. He claimed there was a possibility that I would choose to break my promise; I claimed there wasn't. So clearly he was right and I was wrong.

Determinist: I would not put it quite that way. You were wrong, of course. But I would not say he was right. He was wrong too.

Victim: What do you mean?

Determinist: He was right in that it was not a certainty that you would keep your promise—this you now realize. But he was wrong in saying that *it was up to you* whether or not you kept your promise. His basic fallacy is that he believes in free will.

Victim: That's no fallacy! I also believe in free will.

Determinist: You mean you *used* to believe in free will.

Victim: No, I still believe in free will.

Determinist: Even after your experience with the brain surgeon? Even after he forced you to break your promise?

Victim: I was not *forced* to break my promise; I *chose* to break my promise.

Determinist: You really mean to say that you actually believe that it was *your* choice in the matter rather than the surgery which caused you to break your promise?

Victim: Of course it was my choice!

Determinist: You sound quite proud of the fact!

Victim: Hardly! How could anyone in his right mind be proud of breaking a promise?

Determinist: Well then, let me ask you: Why *did* you break your promise?

Victim: Why did I? Because I realized that it was the best thing I could do, that's why?

Determinist: But do you believe it is morally *right* to break a promise?

Victim: Of course not! But in this case, my keeping the promise would have been even worse than breaking it. I thus had to choose the lesser of two wrongs.

Determinist: I don't understand you.

Victim: Look, when I originally made the promise, I did so in good faith. I honestly believed it was best for all concerned. But shortly after my brain operation, I started to review all relevant facts, and I soon discovered some of the more crucial ones I had failed to take into consideration. After having taken them into consideration, I then realized that the promise was not a good one—I don't mean not good for me personally, but not good for everyone as a whole. In other words, I now realize that keeping the promise would have been even more unethical than breaking it—despite the fact that I fully grant that it is always unethical to break a promise. But unfortunately, in life the choice is not always between the ethical and the unethical, but sometimes between the more unethical and the less unethical. Whenever someone makes an unethical promise, he is always confronted with this horrible choice. If he breaks his promise, he is committing the unethical fact of breaking a promise; if he keeps his promise, he may, for other reasons, be acting even more unethically. This was my situation. I had no choice but to break my promise.

Determinist: You sure as hell had no choice! The brain surgeon saw to that!

Victim: No, no; I mean I had a choice, and I chose to do the best thing under the circumstances.

Determinist: Now whom do you think you are kidding? Look how you rationalize and flounder! Next, I suppose, you'll be telling me that the brain operation had nothing to do with your so-called "choice."

Victim: No, I'm not saying *that*. It is quite possible that the brain operation set in motion various thoughts concerning the desirability of keeping my promise. Had I not had these thoughts, I would indeed have kept my promise. But the brain operation did not *force* me to break my promise; I broke it of my own accord. I *chose* to break my promise.

Determinist: And I say that that is a complete and utter rationalization! Of course the brain operation caused you to break your promise! If you had not had the brain operation, you would have kept your promise, but since you did have the operation, you broke it. It is as simple as that! If event X is followed by event Y, and if Y would not have occurred had X not occurred, how can you intelligently deny that X was the cause of Y? What other meaning of "cause" is there?

Victim: I disagree! It is inconceivable to me that I did not break my promise of my own free will. When I make a promise, there is no power in the universe—other than my own will—which can cause me to break it.

Schizophrenic Seminar

☐ This is a conversation between four schizophrenic friends in a lunatic asylum in moments of exceptional lucidity.

First Schizophrenic: It might interest you to know *how* I became psychotic. All my life I have been an atheist. I have always known that there is no God in this world; the idea that there *is* is obviously only a wishful superstition. But my whole being revolted against a Godless world, hence I have managed to escape from this real world in which there is no God into my own private world of fantasy in which there *is* a God. And in this fantasy world, I am *so* happy!

Second Schizophrenic: That is interesting; my case is virtually the opposite! All my life I have known that there *must* be a God in this world. But I have always found the thought of God so infinitely frightening and nightmarish that I would have killed myself long ago to escape from this hideous reality, except for my fear of eternal punishment. So instead of this physical escape I have taken the "psychological" escape into a private fantasy world in which God does not exist. And in this world, I am so happy!

Third Schizophrenic: How ironical! From my point of view, the statement "there is a God" is a sheer fantasy, and the statement "there isn't a God" is just as much of a fantasy— there is no way one can verify either. So although you each *think* you have escaped from the real world, you have merely escaped from one fantasy world into another. Indeed, you have merely exchanged each other's fantasy worlds!

Now, my case is different. I believe not only that each of the statements "there is a God," "there isn't a God" is a fantasy, but also that it is a fantasy to assert that either there is a God or there isn't. I say this because I identify truth with verifiability, and since neither component of the disjunction is verifiable, neither one can be true, hence the disjunction is not true. But according to the law of the excluded middle, which is a famous law of classical logic, any proposition either is true or it isn't. This means that classical logic does not apply to the real world. But I found it so insecure living in a world going counter to classical logic, that I had to escape into a fantasy world in which all classical laws of logic apply. And in this world, I am very happy.

Fourth Schizophrenic: I find your case the most ironical of all! You *think* you have escaped from the real world into a fantasy world, but in reality you have done the very reverse! Look, the law of the excluded middle obviously *is* sound and *does* hold for the real world—anybody in his right mind knows that! So when you doubted the law of the excluded middle, *then* you were living in a fantasy world. But now that the law does apply to your present world, your present world is the real one. Indeed, I am surprised that you are still with us!

Now, *my* case is very, *very* different! I have had many reasons to wish to escape from the real world to a fantasy world. But I could see no way actually to do this! The problem got increasingly insurmountable, until I suddenly realized the following: Although I vehemently affirm the law of the excluded middle, I just as vehemently deny the law of contradiction. I certainly believe that some propositions are both true and false. In particular, this world is both the real world and a fantasy world. So to escape, I had merely to remain where I was!

An Amazing Christian Scientist

☐ Once there was a person who was both a Christian Scientist and an excellent logician. We will call him "Christus," or "Chris" for short. Chris's views were perhaps not wholly orthodox. Some said he was not a *true* Christian Scientist, but he maintained he was. He said he was by virtue of the fact that he wholeheartedly assented to the following four basic propositions: (1) matter is unreal; (2) evil is unreal; (3) suffering is unreal; (4) faith can cure all. With enough faith, one does not need any doctors.

One day Chris had a severe infection. He went to a doctor

to get an antibiotic. On the way he met a friend, a fellow Christian Scientist, who realized his intentions. The following little dialogue ensued:

Friend: How come you are seeing a doctor? I thought you subscribed to the doctrines of Christian Science.
Chris: I do.
Friend: Then why are you seeing a doctor?
Chris: You believe in Christian Science, don't you?
Friend: Of course!
Chris: Then you agree that matter is unreal?
Friend: Of course!
Chris: In that case, my body is unreal, the doctor's body is unreal, his office and medical instruments and drugs are all unreal. In what sense then am I seeing a doctor?
Friend (after a moment of dazed silence): Good God, what sort of insane sophistry is this! You must be kidding!
Chris: I am not kidding, and this is not sophistry. It is plain cold logic!
Friend: If that's what you call "logic," I'm afraid I don't have much use for your "logic"! Look, now, are you *seriously* maintaining that you are not seeing a doctor?
Chris: How could I be? Doctors are unreal!
Friend: But this is insanity!
Chris: More so than our belief that matter is unreal?
Friend: Look, I cannot take you seriously! I suspect that you are in our church only to *sabotage* the cause of Christian Science. No true Christian Scientist would talk like this!
Chris: But he *must*, if he is to be consistent!
Friend: You are not being *consistent*; you are, I fear, only being hypocritical. You call yourself a Christian Scientist, yet you run to a doctor because of your lack of real faith. And then you rationalize your acts by denying that you are seeing a doctor, and you call doctors "unreal"!
Chris: But they are! I grant it may *appear* as if I am seeing a

doctor, but my seeing a doctor is true only in the world of appearance, not in the world of reality! It holds in the phenomenal but not in the noumenal world.

Friend: I don't know much about philosophy and "noumena" or "phenomena." I am a simple, honest Christian Scientist, and I cannot believe a word of what you are saying. I cannot believe that you really believe that you are not seeing a doctor.

Chris: But of course I'm not! Since I am committed to the proposition that matter is unreal, then I *must* believe that my seeing a doctor—though it *appears* to be true—must *in reality* be false.

I'm afraid I must totally agree with Chris! If matter is unreal, so are doctors—it is as simple as that.

Now comes a more subtle point: The friend walked away from Chris in disgust. Then Chris met a second Christian Science friend, and the following dialogue took place:

Friend: How come you are visiting a doctor? I thought you believed in Christian Science.

Chris: I do.

Friend: Then you must believe that enough faith can heal.

Chris: I do.

Friend: Then why do you see a doctor?

Chris: Because I don't have enough faith.

Friend: Oh, then you *don't* fully believe in Christian Science.

Chris: Yes, I do.

Friend: I am puzzled! Explain!

Chris: I believe that complete faith *is* adequate for healing. If I had enough faith that I could get well, I would not need to go to a doctor. But I am realistic enough to know that I probably do not have enough faith for this. Hence I will see a doctor.

Friend: Oh, so you *don't* have complete faith in Christian Science!

Chris: No, you totally misunderstand me! I do have complete faith in the *doctrines* of Christian Science. Specifically, I have complete faith in the fact that *if I had enough faith that I would get well without medical help, then I would.* In this fact I have *complete* faith, and therefore I feel justified in calling myself a Christian Scientist.

Friend: Then why do you see a doctor?

Chris: Because I *don't* have enough faith that I will get well without one. All I have faith in is that *if* I had enough faith that I would get well, then I would. Remember now! I do not have faith that *everybody* gets well—nor do you. All that Christian Science teaches us is that those *with enough faith* get well. You know as well as I do that many Christian Scientists have refused medical help and as a result have died.

Friend: Death is unreal!

Chris: Oh come on now, on a metaphysical level you are right, but don't you honestly know what I meant when I said "they died"?

Friend: Yes, I guess so. But then they died only because they didn't have enough faith!

Chris: Exactly! And that is the whole point! I don't want to be one of those who died for "lack of faith." I want realistically to realize my limitations. Just because I realize that *if* I had enough faith I would get well, how do I know that I *do* have enough faith to get well? Do *you* know whether or not I do?

Friend: Of course not! Only *you* can know that. However, don't you realize that by going to a doctor, you are only weakening your faith? If you stay away from a doctor, then you will strengthen your faith.

Chris: I believe that! I really do! All right, if I don't see a doctor, then I grant that I will have more faith than I do now. But will I then have enough faith to get well?

Friend: How can I know that?

Chris: You mean you have doubts?

Friend: Of course!

Chris: Good God! Then how can you recommend that I stay away from a doctor? If you felt *sure* I would get well without a doctor, then it would make perfect sense for you to recommend that I don't see one. But you now realize I am right in that if I stay away from a doctor, my recovery is highly problematical. I will then get well only if I have enough faith. But you don't know if I will have enough faith, hence you don't know that I will get well without medical help. How then can you possibly take the responsibility of advising me not to see a doctor?

Friend (after a long pause): I guess you are right; I will have to retract what I said. I see now that if I do not feel confident that you will have enough faith, then I cannot in good faith advise you to stay away from a doctor. All right then, I must retract what I said about having "doubts" that you won't have enough faith. So now I say, "Yes, I believe if you don't see a doctor, then you will have enough faith to get well without one."

Chris: How can you know this?

Friend: I have *faith* in it.

Chris: Wouldn't you say that to everyone in my position?

Friend: Of course!

Chris: Then you would have said that also to those Christian Scientists who stayed away from doctors and who died as a result of lack of sufficient faith.

Friend: I guess so.

Chris: But in these cases you would have been wrong.

Friend: I suppose so. Yet in these cases, one cannot be sure. Had I—and enough others like me—been there to advise them, and had we manifested our faith in their faith, our faith might have reinforced their faith to the point that they wouldn't have died. We strongly believe that faith reinforces faith.

Chris: Well now, suppose I took your advice and stayed away from a doctor, and subsequently died. How would you feel?

Friend: That you didn't have enough faith.

Chris (irritated): Yes, yes, we know that! We have already agreed about that! The point, though, is that if I stay away from a doctor and die as a result of lack of faith, then would you not realize you had been wrong in believing that I had enough faith to get well without a doctor?

Friend: Yes, I guess I would have to!

Chris: Then you would have given me bad advice! Wouldn't you then feel partly *responsible* for my death? Would you feel no guilt whatsoever? Wouldn't you perhaps feel, "Good God, if only I hadn't advised him to act against his better judgment, he might still be with us today!"?

Friend (after a long pause): You certainly put me on the spot! I don't know what to say! Logically, I guess you are right. Yet something deeper within me bids me act as I do. The whole point is that one should *not* compromise with lack of faith. Granted that faith sometimes falters, still in the long run it is best that we have faith, and also faith that our faith won't falter, and so forth. I can see many possible objections to this whole point of view, yet with enough faith, these very objections lose their emotive power.

This concludes my story. Oddly enough, I can appreciate and sympathize with both points of view. Logically, I would say Chris is in the right. If I were a Christian Scientist (which is not likely, but not impossible), I would believe and act as Chris did. The belief that enough faith can cure without medical help does *not* imply that I should stay away from doctors. On the other hand, I understand the friend's point that it is not good to compromise with lack of faith.

Let me now tell you my attitudes toward Christian Science. The worst thing about it, of course, is its casualties. Particularly unfortunate is the fact that children, who have no say in the matter, die for lack of medical attention. A case was once brought to court, and the court *forced* the parents to give the child medical attention. Had I been this judge, I would have concurred heartily! If you claim I am being intolerant, and ask me "Doesn't a person have the right to practice his own religion?" my answer is *no*. Of course a person has the right to go to the church of his choice, to think as he wants, to say and write what he wants, but certainly not to *act* as he wants just because his religion sanctions or even demands the action. At any rate, I fully share the objection of opponents of Christian Science on the issue of casualties. However, this is only one side of the story!

To evaluate Christian Science fairly, we must also investigate how many psychosomatic ailments, which might be fatal, have been cured by some form of "faith." The situation is not so simple! True, the Christian Scientists probably see doctors less frequently than do others, but is their *need* to see them as great? This is not something which can be answered *a priori*; some empirical investigation is necessary. The crucial question is what is the *overall* effect of Christian Science on health and longevity? And for this, statistical data may be most helpful. What *is* the average life span and general state of health of the Christian Scientist compared with the non-Christian Scientist?

There is still more to the picture! Take some of the propositions believed by Christian Scientists. What about the proposition that matter is unreal? Frankly, I don't know whether matter is real or not (if you want to know whether or not matter is real, perhaps you'd better ask a philosopher!). What about suffering? Well, suffering I know to be real, since I have experienced it.[1] But even if suffering is real, I'm not sure that it's such a bad idea to believe it isn't! Such a belief may, in a strange way, be sort of self-fulfilling, that is, it is possible that the belief in the ultimate unreality of suffering may in the long run tend to reduce suffering. I say this is *possible*; I do not assert that it is necessarily true. I have more confidence that this may hold with faith healing. Frankly, I find it psychologically impossible to believe that suffering is unreal, but I do not

[1]The case of matter is somewhat different. It is questionable whether we experience matter directly, or whether we merely experience certain sensations and bring in matter as a "hypothesis" to account for these sensations.

find it impossible to believe that faith can cure suffering. Again, I am not saying that faith *can* cure suffering, but only that it seems to me a genuine possibility that it can. As for the problem of "evil," here I come extremely close to the attitudes of Christian Science! Evil I believe is sort of unreal, in a way, and what reality it does have is only strengthened by our belief in it and our opposition to it. I feel especially strongly that "moral condemnation" of evil is one of the things which most strengthen what evil there is.

To summarize my position on Christian Science with respect to the question of suffering, I believe suffering is real and the Christian Scientists are wrong when they say it is not. But this wrong belief may be a good one to have, and may have some partial truth (of a self-fulfilling nature) since the *belief* in the nonreality of suffering may well reduce the sufferings of the world. In short, I am not so sure that if everyone were a Christian Scientist, the world might not be a much happier place. This belief of mine is not enough to make me a Christian Scientist, because I am not that pragmatic at heart. That is to say, I am not the type who can believe something just because I believe it is *helpful* to believe it. But I can well understand those who are constituted differently in this respect. In short, if you gave me a choice between a Christian Scientist and one who is highly intolerant of Christian Science, I would prefer the Christian Scientist.

I also evaluate movements in terms of what I believe to be their future potentialities rather than just their past and present performances. And Christian Science may develop along very nice lines. I believe it is here to stay for quite a long while. But I believe it will drastically reform. I predict that in a hundred years there will still be Christian Science churches, but the members will be more like "Chris" and will give up the requirement not to see doctors. I think they will say something like this: "Of course, medical treatment *as a purely physical process* is of no help, but if seeing a doctor gives you more *faith* that you will get well, then by all means see a doctor!"

V DIFFICULT TO PLAY WELL!

A Remark on Spontaneity

☐ I have been told—by learned sources—that when a truly great musician plays, and his playing sounds so free and *spontaneous*, that the listener has no idea of the amount of nonspontaneous work, study, scholarship, analysis, and "planning" required to achieve these spontaneous effects.

I shall not argue the point. I only wish to say that if it is true, it leads me to the amazing realization that spontaneity does not come by itself!

This Is Very Difficult to Play Well!

☐ A rather influential piano teacher once told me that she found the Schubert piano sonatas very "difficult" to play well: their simplicity is deceptive; there are really many musical "problems" in them. Many people have complained to me that it is very difficult to play Bach "musically."

Now, I do believe that it is indeed true that there are very few people in the world these days—compared with the nineteenth century, or indeed even thirty years ago—who play Bach or Schubert musically (or anyone else, for that matter!). But to say that it is "difficult" to play musically is the most misleading thing in the world!

Let me give you an excellent analogy (which many readers will say is a very *bad* analogy). Is it difficult for a bird to sing like a nightingale? It is indeed *rare* for a bird to sing like a nightingale. The only birds who sing like nightingales are nightingales, and among birds, there are comparatively few nightingales (or am I wrong?). Therefore it is *uncommon* for a bird to sing like a nightingale. But it is completely wrong to say that it is *difficult* for a bird to sing like a nightingale. If the bird happens to be a nightingale, then it is the easiest and most natural thing in the world for it to sing like a nightingale. If it is not a nightingale, then it is not that it is *difficult* for it to sing like a nightingale, it is simply *impossible* for it to sing like a nightingale! Any non-nightingale who would try to sing like a nightingale would only make a fool of itself. Fortunately, birds, unlike people, are too sensible to try anything that ridiculous. Now, it *might* happen that a bird who is not a nightingale, and who has not been brought up with the idea that it should be "original" or that "copying" another bird would be beneath its "dignity"—such a bird might indeed hear a nightingale singing, become enchanted with the beauty of its song, and never quite sing the same again. It might then sing *more* like a nightingale. But this is very different from "trying" to sing like a nightingale, or saying "how *difficult* it is to sing like a nightingale."

Many will of course say that my analogy is terrible. After all, they will say, the music of Bach or Schubert is *not* like the singing of a nightingale; it is a highly refined, cultivated, disciplined, and *civilized* art. It requires *thought* and *practice* and *work*! Of course it takes longer to play classical music than to sing like a nightingale, but does this mean that it is necessarily any different in *spirit* from the singing of a nightingale? I'm afraid that most professional musicians today, who are essentially artisans rather than true artists, play in a manner *very* different from the singing of a nightingale. But when a true artist like Edwin Fischer played, it was *exactly* like the singing of a nightingale. And it is this quality which so many of our contemporary artists have lost!

Now, let me get back to the main point. I believe there is no such thing as a musical "problem." Either one's temperament is in accord with a given composer, or it isn't. The piano teacher I first told you about will *never* play Schubert beautifully (at least not in this life!). It's not that she is lacking in "gifts" or "talents" or anything like that; it's just that her personality is about as non-Schubertian as it can be. I'm not saying that she was *born* this way (I believe everybody is born more or less Schubertian, but society soon knocks it out of one), but *now* she is this way. Yes, if her whole life changed radically, if she lost some of her crazy inhibitions, if her true unborn Schubertian nature were allowed to grow, *then* she might play Schubert beautifully. But she never will as a result of any effort to solve "musical problems."

Let me conclude this essay with two jokes. The first is about a man in a subway seeing, to his utter amazement, another man reading a Jewish newspaper upside down. He said, "How can you read a Jewish newspaper upside down?" The other answered, "Do you think it's easy?"

The other story is actually a true one. Johnson and Boswell were together at a concert in which some violin virtuoso had just sweated through a very difficult piece. Boswell said, "That piece must have been very difficult." Johnson answered, "Difficult? I wish it had been impossible!"

Problems

☐ Once when I was playing for a musician, he complimented me on the way I played a particular passage. He told me how well I handled a certain modulation and added, "You don't realize in what a remarkable way you have solved this problem!"

I must say, I was thunderstruck! In the first place, I was not even aware that there was a modulation. (That shows how much *I* know about music! I never think in terms of modulations. I do not deny that they exist; I just don't think about them.) In the second place, I was totally unaware of any *problem* let alone *solving* one! The whole idea of "problem solving," especially in music, strikes me as so weird! Not only weird, but most disharmonious and destructive. Is that how you think of life, as a series of *problems* to be solved? No wonder you don't enjoy living more than you do!

To compliment a musician, or any other artist, on having "solved problems" is to me absolutely analogous to complimenting the waves of the ocean for solving such a complex system of partial differential equations. Of course the ocean does its "waving" in accordance with these differential equations, but it hardly solves them. I do not claim to know whether the ocean is or is not a conscious being, but if the ocean does think (which wouldn't surprise me), the one thing I'm sure the ocean does *not* think about is differential equations.

Perhaps I am allergic to the word "problem." If so, I am grateful for this allergy. Some of you will say I am only quibbling about words. This is not so. It is *ideas* that count, not words. And I believe that one who feels he is "solving problems" lives very differently from one who does not feel this way. I believe my objection to the notion of "problem" is due to my deep conviction that the moment one labels something as a "problem," that's when the real problem starts.

Bees and Scholars

☐ When I was young, I was intensely attracted to a very few fields, and only to very limited areas within those fields. I had absolutely no patience with the notion of a "general, well-rounded education," and I was (and still am) extremely resentful that schools tried to force a general education upon me rather than allowing me to pursue my own interests. A close friend of mine at the time (a man ten years older than I) who was a staunch defender of the existing school system—with all its horrid battery of "required courses"—took a very disdainful view of my attitude. He said, "As for me, I have all my life had the ideal of being a cultured gentleman." Now, I do not object to one's being a cultured gentleman, though the *concept* "cultured gentleman" strikes me as less nice than an *actual* cultured gentlemen. And I find the *idea* of being a cultured gentleman somewhat repellent. There is nothing wrong with being a cultured gentleman, provided the "culture" comes as a by-product of the pursuit of one's esthetic or intellectual passions, rather than something to be pursued directly. I feel much the same about so-called "scholarship." If a person has a healthy appetite for a field, he will in time acquire a good knowledge of the field as a matter of course; he does not have to set out deliberately to be a "scholar" in the field. It is the difference between what I might term the "natural" scholar and the "studied" or "self-conscious" scholar. I still haven't said it quite right; let me try again.

I love the artist or scholar whose activity is like the bee pursuing the delicious nectar of the flowers. The bee has no mind to become a renowned authority on which flowers contain the best nectar; the bee simply loves nectar. In all probability, the bee, through his actual experience will soon have a *fantastic* knowledge of the flower geography of his neighborhood—as good perhaps as any human scholar who "studies" botany. And I say the bee really knows the flower much better than the botanist. The botanist merely knows *about* the flower; the bee knows the flower directly. The more analytically minded reader might well ask, at this point, just what I mean by "knowing *about* something" versus "knowing it *directly*." I wish I could answer him! The distinction is so difficult to explain rationally, and yet it is of such vital importance.

To return to my main point: what I am trying to say, of course, is that I love the approach of the bee rather than the approach of the self-conscious scholar. And I again emphasize that I do not object to scholarship providing it comes as a mere by-product of the pursuit of nectar. Of course, it may be that to some, scholarship itself is a kind of nectar. To which I can only say, "to each his own." Indeed, it may even be that the very dry pedantic scholar is, in his own weird way, pursuing his own very dry kind of nectar.

Four Types of Life

☐ I shall illustrate my point by considering musicians, though what I have to say is equally applicable to other walks of life.

The first type is the man who is always complaining that he is not "getting anywhere." He says that he does not learn fast enough, that his fingers are very bad that day, that his memory is poor, that he is not "with the music," that things are too "mechanical," that he does not practice enough, that he doesn't have enough "self-discipline" to force himself to go through all the gruesome, boring, painful details, that he knows that discipline and self-denial are necessary for the "perfection of an art," but that he doesn't have enough of these qualities, etc. And so he complains and gripes and complains and gripes, but through all this restless uneasiness, he progresses anyhow and eventually becomes a first-rate musician. After he is first rate, he may continue to complain and gripe, but this does not prevent him from being first rate. This concludes my description of the first type.

Now for the second type: He is virtually the opposite. He is with his instrument—say the piano—for hours and hours a day, he is in a state of complete ecstasy, he just enjoys himself at the piano playing away merrily like a bird sings, he does not distinguish between "playing" and "practicing," he has no conscious idea of "improving," he does not "try to play well," he has no idea whether he is playing well or playing badly, and he couldn't care less—indeed, he never listens to himself playing, but only to the music that he is playing (just as you and I usually read a book without worrying whether we are "good readers," and with no thought that we are "progressing" or "learning a skill"). That's the whole point: our second type approaches music without any thought that he is "learning." He is not "teaching himself," he plays but is not consciously "learning to play." And so he plays merrily away all the hours of the day he can find, and never complains because he has nothing to complain about; he is enjoying himself too much! However, he never progresses! Or to be more realistic, let us say that he automatically progresses somewhat, but not in a way commensurate with his talents. Let us say that he has the capacity for becoming first rate, but because of his approach he never does become first rate. People say to him of his playing: "It is beautiful and moving. You have the true soul of an artist. At moments it is as beautiful as beautiful can be. You sound as if you so love and *enjoy* music. Certain passages in isolation you play as beautifully as they can be played. But the composition as a whole lacks cohesion, unity, depth, architecture, form, and discipline. It is sad, you have first-rate talent and could be first rate if only you would seriously apply yourself and use more self-discipline." This concludes my description of the second type.

Well, dear reader, which of the two types do you prefer? Or perhaps I should ask which of the two types do you find more horrible? Speaking for myself, I find the first type a gruesome nightmare! I don't care *how* good he finally gets, to me he is a gruesome nightmare. Some may find great beauty, nobility, and heroism in his willingness to endure so much dissatisfaction for the sake of his "art," but I still find such a person a gruesome nightmare. As for the second type, I find him not a gruesome nightmare, but rather sad. It is obviously sad that anyone should fail to develop his full capacities. And so I find the second type sad. Must we choose between being of the first or of the second type? I shall not argue the point here. I merely wish to say that if that is the best choice life has to offer, then I wish to offer my condolences to the universe.

Now I wish to consider two more types, each of which combines certain features of the first and second types.

The third type combines all the disadvantages of the first and second. This type grumbles and complains away, finds practicing very tedious and unpleasant, is constantly regretting his lack of determination and lack of self-discipline, but also gets nowhere. After a few years of this frustrating activity, he is likely to give up his music altogether. He says things like: "It is because my standards are *so high* that I have given up playing myself. It is pointless to play just like an amateur. Music is a great thing and should be attempted only by those with more talent and perseverance than I possess." Yes, I actually know people with this attitude! They also have a tendency to be mercilessly critical toward those who *do* keep up their own playing. They spend all their time and energies criticizing others rather than developing themselves. Well, what do you think of *this* type? I guess everyone will find this type distasteful.

The fourth type (if he exists) combines all the advantages of the first and second types. His approach is very much like that of the second type; he just enjoys himself for hours and hours a day playing merrily away as spontaneously as a bird sings. He has no grim sense of "duty" or "discipline" or "sacrifice," he just has an enormous love of beauty and pursues it relentlessly. He may have a great critical faculty, but entirely on an unconscious level. He may, for example, play a particular passage over and over and over again, but he does not think of it as "practicing" or "learning." He does indeed sift and sort, but he does not know that he is "sifting and sorting." He is much like a dog who is offered a dish containing a mixture of good foods and bad foods. He spits out the bad foods and eats up the good foods. The doggie does not complain nor criticize the bad foods; he is far too busy enjoying himself hunting out the good foods. And so after the meal is over, if someone asked him how he enjoyed his dinner, the doggie would say, "Delicious! I found so many good foods!" (By contrast, the one of the first type would say, "Horrible! Most of the food was bad! I had a hell of a time finding enough good food to make a meal.") And so our artist of the fourth type is like our doggie; he has a sensitive, discriminating snout and hence sorts and sifts out the good, and bothers so little with the bad that he scarcely remembers it—he just rejects it automatically. So therefore this type is not so much "consciously critical" as "unconsciously selective and discriminating." The behavior of this type often fools other people completely. People hearing him practicing may say: "This fellow is amazing! He has so much patience! He spends hours and hours on details! He has so much self-discipline! He does not balk at performing irksome tasks. He is obviously highly self-critical, and that's why he improves! He does not just *take* what his fingers give him, he *tells* his fingers what to do, he masters his fingers rather than allowing his fingers to master him. He schools himself, he disciplines himself, he overcomes obstacles, and that's why he succeeds. He listens to himself and corrects himself." Yes, that's what people say of him, but they are totally wrong! In reality, this man no more "evaluates himself" than the doggie selecting the good foods evaluates his own performance as a "food selector." Of course the doggie becomes better and better as a food selector with more and more experience, but the doggie is not training himself to be a "professional food selector." The doggie is not conscious of himself at all, but only of the foods he is selecting. And likewise our artist is not trying to improve himself, but rather is automatically selecting those ways of playing which produce the most beautiful results. Of course, our fourth type ultimately becomes a first-rate musician just like our first type, but what a difference of approach! The first type has the ambition of becoming a "great food selector," whereas the fourth type simply loves good food.

To Know Is Not as Good as to Enjoy

☐ One of my favorite sayings of Confucius is about education. He said, "To know is not as good as to love. To love is not as good as to enjoy."

Spanning fourteen centuries, we find the typical eighth-century Chinese eremitic poem going something like this.

My friend comes to visit me from across the hills.
We fill our cups and enjoy the classics together.

The delightful thing is that the poet would say "enjoy" rather than "study" or "learn." Evidently the classics were something to be loved and enjoyed rather than "studied" or "learned."

There we have the true Taoist concept of learning! Fortunately this ideal is not confined to the ancient Chinese. Spanning eleven more centuries, and several thousand miles westward, we find the father of Robert Browning displaying a similar attitude toward the education of his son. The following passage is from William Lyon Phelps's book _Robert Browning._[1]

In these days, when there is such a strong reaction everywhere against the elective system in education, it is interesting to remember that Browning's education was simply the elective system pushed to its last possibility. It is perhaps safe to say that no learned man in modern times ever had so little of school and college. His education depended absolutely and exclusively on his inclinations; he was encouraged to study anything he wished. His father granted him perfect liberty, never sent him to any "institution of learning" and allowed him to do exactly as he chose, simply providing competent instruction in whatever subject the youth expressed any interest. Thus he learned Greek, Latin, the modern languages, music (harmony and counterpoint, as well as piano and organ), chemistry (a private laboratory was fitted up in the house), history and art. Now every one knows that so far as definite acquisition of knowledge is concerned, our schools and colleges—at least in America—leave much to be desired; our boys and girls study the classics for years without being able to read a page at sight; and the modern languages show a similarly meager harvest. If one wants positive and practical results one must employ a private tutor, or work alone in secret.[2]

[1]Indianapolis: The Bobbs Merrill Company, Publishers, 1915.
[2][I must confess, I am puzzled by the clause "or work alone in secret." What on earth did Phelps mean?]

The great advantages of our schools and colleges—except in so far as they inspire intellectual curiosity—are not primarily of a scholarly nature; their strength lies in other directions. The result of Browning's education was that at the age of twenty he knew more than most college graduates ever know; and his knowledge was at his full command. His favorite reading on the train, for example, was a Greek play; one of the reasons why his poetry seems so pedantic is simply because he never realized how ignorant most of us really are. I suppose he did not believe that men could pass years in school and university training and know so little.

This passage interested me enormously; it is an absolutely perfect expression of my own philosophy of education. Many criticize this as being too "elitist." They say, "This may be all right for Robert Browning's father, who had plenty of money, or for Robert Browning, who was doubtless an unusual child. But this method of education would hardly be applicable to the rough and tumble masses." Unfortunately even William Lyon Phelps kind of spoils the above passage by continuing: "Yet the truth is, that most boys, brought up as Browning was, would be utterly unfitted for the active duties and struggles of life, and indeed for the amenities of social intercourse. With ninety-nine out of a hundred, such an education, so far as it made for either happiness or efficiency, would be a failure. But Browning was the hundredth man. He was profoundly learned without pedantry and without conceit; and he was primarily a social being."

I am sad that Phelps believed that Browning's kind of education is inapplicable to the masses. *This* attitude strikes me as elitist! Why cannot schools—even, or perhaps particularly, elementary schools—have teachers of the spiritual caliber of Robert Browning's father? I am thinking now of elementary schools *whose teachers have a Ph.D. or the equivalent and preferably are world outstanding scholars*. Yes, I am suggesting the radical proposal that the intellectual cream of the crop should not be entirely wasted at universities and graduate schools, but should be at high schools and elementary schools where they are even more sorely needed. This would mean, of course, that their teaching hours would be no greater than at universities, and their salaries no lower. This might sound very expensive! But what better way could a nation spend its money than on the *true* education of its young? Some may think that an outstanding world scholar may be *wasted* at an elementary school. But this is wrong! Children recognize and appreciate true quality far better than many adults realize.

To avoid misunderstanding, the last thing in the world that I would want is that professors be *forced* to teach at elementary or high schools! Each professor should teach where he wants. I would like to see only that elementary and high school positions be made as attractive as those at universities. Ideally, I would envisage elementary schools, high schools, colleges, and graduate schools all on the very same campus, with a common teaching staff, where it might be quite common for a professor to be simultaneously teaching one course at a college or graduate level and another at a high school or elementary school level. I think it would be marvelous for youngsters to have frequent contact with college and graduate students.

Needless to say, I have all my life been against examinations and required courses. Concerning examinations, I am not primarily objecting to an employer who wishes to examine a candidate for his competency for the job, nor to any institution of higher learning that wishes to examine *incoming* students for their qualifications to pursue their studies. What I really *am* objecting to—and quite strongly at that—is the necessity for the teacher of a course to have to report to others the progress of the student. To report a student's failure seems to me like an utter betrayal! It is almost like breaking the Hippocratic oath. And it certainly destroys any decent relationship between teacher and student; it puts the teacher in the role of an adversary rather than a friend.

As to required courses, the usual argument in their favor is, of course, that students are too immature to know what is "good for them" and that they should be "exposed" to things which may later be found to be of interest or use. I have known advocates of required courses who have said things like, "I'm *glad* that I was forced to learn so and so; if I hadn't been, I would never have come to it by myself." Now, it so happens that my own personal experiences in life have led me to the utter opposite point of view. I can honestly testify that I have never benefited one iota from anything I have ever been required to learn; all I got from all this was pain, frustration, heartache, and bitterness.

The views of Albert Einstein toward his own formal education deserve to be far more widely known than they appear to be. He found the coercive process of being forced to cram all sorts of stuff into one's mind, whether one liked it or not, to have had such a deterring effect, that after passing his final examination, he could not bear to consider any scientific problem for over a year! He compares this process to robbing a perfectly healthy beast of its voraciousness by forcing it to devour continuously, even when not hungry. Einstein held it a grave mistake to believe that the enjoyment of seeing and seeking can be promoted by means of coercion and a sense of duty. Indeed, he regarded it as short of a miracle that the traditional methods of formal education had not entirely strangled the holy curiosity of inquiry.

Of course some readers will say: "Yes, that attitude is all right for Einstein, but we can't all be Einsteins. Maybe for *him* this coercive educational policy was not good, but for us ordinary people, without this coercion, we wouldn't learn." To which I wish to reply: Has it never occurred to you that this attitude of Einstein may itself be one of the contributing factors of his greatness?

Postscript: The following views of Emerson on college education also deserve to be more widely known than they apparently are.

If the colleges were better, if they really had it, you would need to get the police at the gates to keep order in the in-rushing multitude.—See in college how we thwart the natural love of learning by leaving the natural method of teaching what each wishes to learn, and insisting that you shall learn what you have no taste or capacity for.—The college, which should be a place of delightful labor, is made odious and unhealthy, and the young men are tempted to frivolous amusements to rally their jaded spirits.—I would have the studies elective. Scholarship is to be created not by compulsion, but by awakening a pure interest in knowledge. The wise instructor accomplishes this by opening to his pupils precisely the attractions the study has for himself.—The marking is a system for schools, not for the college; for boys, not for men; and it is an ungracious work to put on a professor.[3]

[3]See James Elliot Cabot, *A Memoir of Ralph Waldo Emerson* (Boston and New York: Houghton, Mifflin Company, The Riverside Press, Cambridge, 1887), Vol. 2, pp. 614–19, 631.

VI IS ZEN PARADOXICAL?

The Sage and the Dog

☐ I wonder if there is in the ancient Chinese literature anything to the following effect:

"The Sage and the Dog Are Indistinguishable"

Of course, one might reply that dogs have four legs and sages only two (or to be more precise, the average dog has four legs and the average sage two). But this distinction is quite superficial compared with the profound similarity of the sage and the dog. What is this profound similarity? Unfortunately I cannot tell you, since I am neither a sage nor a dog.

A Zen Fragment

☐ I read in a Zen sermon:

"Today I wish to talk to you about 'nondoing.' Nondoing means that there is nothing to be done and nothing to be learned. So I must first talk to you about Nothing. But alas! Any words I use to talk about nothing are so apt to miss the point!"

What Is Zen?

☐ If you don't know what Zen is, no matter. On a purely rational level, no one knows what Zen is. Indeed, if one *knew* what Zen was, it would cease to be Zen. Trying to define Zen is like trying to define jazz. One cannot know what jazz really is unless one has heard it. As one jazz-ist said: "Jazz is a feeling. It's the feeling you get when you know you're going to get a feeling."

This reminds me, somewhat, of the reply given by the Zen philosopher Daisetz Suzuki to the question: How would you characterize the experience of Satori (Zen-enlightenment)? He replied: "It is much like everyday experience, only about three-quarters of an inch off the ground."

Definitions are sometimes strange things! Except in mathematics and the exact sciences, they are apt to miss the point. And even in mathematics (or rather the philosophy of mathematics), Bertrand Russell gave the famous definition: "Mathematics is the subject in which we do not know what we are talking about, nor whether what we are saying is true." The logician Stanley Tennenbaum, in a celebrated invited address, said of this definition: "*The* subject? I would have thought there were others!" Well, Zen might be one such subject. Indeed, Zen may well be *the* subject in which one does not know what one is talking about, nor whether what one is saying is true.

A graduate philosophy student, who took a course with me in logic, told me of a course he once had in Zen. He described the proceedings as follows: "On the first day, the teacher came in and started talking a lot of nonsense to the class. The students then answered with complete nonsense. Then the teacher continued talking nonsense; the students again responded with nonsense, and so this went on till the end of the hour. On the second day, exactly the same thing happened. Indeed, this happened day after day until the end of the semester. On the last day, the teacher gave all the students A's."

I find the above incident delightful! Of course, I very much doubt whether the procedure was quite as nonsensical as my student imagined—or if it was, we should bear in mind the keen observation of the philosopher Ludwig Wittgenstein, that certain kinds of nonsense are *important* nonsense.

Is Zen paradoxical? In the last analysis, I believe the answer is *no*. Of course, the entire Zen literature abounds in paradox, but it seems to me that Zen-masters use paradox mainly as a technique to lead one to the state where the entire duality between paradox and nonparadox is transcended.

94

Getting back to the question of trying to define Zen (which, of course, I am not trying to do), in my book *The Tao is Silent*[1] I characterized Zen as a mixture of Chinese Taoism and Indian Buddhism, with a touch of pepper and salt (particularly pepper) thrown in by the Japanese. I think this characterization—particularly the part about the pepper—might give a little of the *flavor* of Zen, but it was hardly intended as a *definition*! It is of the same impressionistic character as Suzuki's characterization of Satori or of the jazz-ists' characterization of jazz.

Let me close with a remark made by Professor Suzuki in a lecture given at Columbia University: "Zen is not as difficult to grasp as you Americans think. It's just that in Zen, we call a spade a nonspade."

A Quintessential Zen Incident

☐ Of all the Zen stories I know, the following captures the quintessence of Zen as well as any.

The story is of a Zen-master who was delivering a sermon to some Zen monks outside his hut. Suddenly he went inside, locked the door, set the hut on fire, and called out, "Unless someone says the right thing, I'm not coming out." Everybody then desperately tried to say the right thing and, of course, failed. Along came a latecomer who wanted to know what all the fuss was about. One of the monks excitedly explained, "The master has locked himself inside and set fire to the hut and unless somebody says the right thing, he won't come out!" Upon which the latecomer said: "Oh my God!" At this the master came out.

I hardly think this story needs any analysis. But for the few readers who might find the story puzzling, a remark or two may not be out of order.

[1]New York: Harper and Row, 1977.

Obviously, what the master wanted was a response which was wholly spontaneous. Any response which was "designed" to persuade the master to come out would have *ipso facto* failed. When the latecomer said, "Oh my God!" he was not "trying to say the right thing"; he did not say it with any idea that it would bring the master out; he was simply alarmed! Had anyone else present said "Oh my God!" for the *purpose* of getting the master to come out, the master would probably have sensed it and accordingly would not have come out.

I also doubt very much that the master had only one definite thing in mind as the "right thing to say." Indeed, I suspect that a purely nonverbal response would have equally satisfied the master. If someone, for example, had tried to break through the window and bodily drag the master out, this might well have been enough for the master to have come out. After all, the master was, in a way, acting like a lunatic (albeit a highly enlightened one), and for the people there just to stand around playing games with a lunatic—and on the lunatic's very own terms!—is precisely the sort of thing the master would wish to discourage.

I still have not been able to say all I see in this very remarkable story. One last thing I will say is that this story illustrates as well as anything can how a situation can arise in which *trying* to bring something about can be the very thing which prevents it.

Humans, Mountains, and Streams

☐ Are human beings like mountains and streams? I'll return to that question later. Meanwhile, let me tell you the following incident: I once made the mistake of telling a social scientist— a very earnest, grim, rather joyless and "rational" person—that I share the point of view of many of the Chinese Taoist artists and poets that human life, at its very best, is like "floating clouds"—sometimes of this shape and sometimes of that. Now this social scientist replied, "You mean human beings are like monkeys?" I was nonplussed by this response, simply because I have had virtually no firsthand experience with monkeys. I have seen photographs and moving pictures of monkeys, and I have seen them a little bit at the zoo, But I have never gotten to really *know* a monkey, in the same sense as I know cats and dogs. You see, with cats and dogs, I know them immediately; I directly experience their Buddha nature. But with monkeys, I have not yet had the privilege of doing this. But if I knew monkeys better, I'm sure I would say, "Yes, monkeys, like humans at their best, are exactly like floating clouds—also like mountains and streams."

This brings us back to the question: "Are human beings like mountains and streams?" This is a profound question, and there seems to be remarkably divergent opinion on the matter! I think many people would say, "Of course they are different!" Especially those brought up in a Jewish or Christian tradition would say: "Humans are alive. They have souls. They are *rational.* They have free will and make *choices.* They are capable of good and evil; they have moral responsibility; they are capable of *sinning*," etc., etc. On the other hand, there are those ultra-mechanistic types—often in the computer sciences—who say that human beings are basically like mountains and streams; they are both causally determined *mechanisms.* The human being (they say) may be a more complicated kind of machine but is a machine nevertheless and subject to exactly the same physical laws which govern the behavior of mountains and streams. Therefore human beings are like mountains and streams.

But what a ghastly, perverted way of looking at it! I don't know which of these two viewpoints I find more horrible! The first emphasizes that aspect of religion I have always hated. The second is completely inhuman. I also tend to think of humans—at their best—as being like mountains and streams, but for such different reasons! In the first place, I don't think of humans as machines. Second, I don't think of the universe as a machine; I think of it as an organism rather than a mechanism. Even if it is perfectly describable, using purely mechanical and electromagnetic laws, I still think of it as an organism rather than a mechanism. And I think of mountains and streams as part of the organism of the cosmos, and a very beautiful part at that. And since I like to think of human beings as also beautiful, I therefore like to think of them—at their best—as being like mountains and streams.

My attitude is very similar to that once expressed by a Korean student of Alfred North Whitehead. In a philosophy course, Whitehead was once explaining the philosophy of naturalistic materialism. At the end of the lecture, the Korean student excitedly came up to Whitehead and exclaimed: "How wonderful! How beautiful! This means that the mountains and streams which I have always so loved are exactly like us!"

Why Does Water Run Downhill?

☐ Supposing I should ask the following question: "Why does water run downhill? Does it *choose* to, or does it *have* to?" I'm sure the vast majority of people today would answer: "It has to. Water is not something alive or conscious, hence it has no power of making choices. Water is something dead and inert. Water runs downhill because it is acted on by the force of gravity; it *has* to do this; it has no choice in the matter."

On the other hand, more (so-called) "primitive" people, who are more animistic in their general philosophical outlook, and who tend to endow mountains and rivers with "spirits" or "souls," would more likely answer that the water *chooses* to run downhill. If asked *why* the water chooses to run downhill, they might be sorely pressed for an answer. Well, suppose I were a primitive animist (which, incidentally, I am) and did not know anything about the law of gravity. I would then believe that water chooses to run downhill, but would be puzzled why. Suppose a physicist came along and told me that it does not choose to, but runs downhill merely because of the law of gravity. I would ask him to explain to me the law of gravity, and he would. Then I would be utterly delighted and say: "Wonderful! Marvelous! Amazing! How utterly beautiful! Now I see why water chooses to run downhill! Of course! The water at the top of the hill is obviously very uncomfortable being subject to this unpleasant force, and hence, to relieve its painful tension, flows downhill where it can be more comfortable and rest in peace. So I was right! Water always seeks the most comfortable places." I wonder if the physicist would feel that I had somehow "missed the point"?

Interestingly enough, if you asked a pure operational scientist, "Does water run downhill because it chooses to or because it has to?" he would answer, "Neither." You see, in science proper there are no such notions as "choosing to" or "having to"; science merely describes what things in fact do happen. So the scientist would say: "It neither chooses to nor has to. It simply *does*." I find it delightful that most Zen Buddhists would agree with him on this point.

A Clod of Earth

☐ Chuangste wrote the following marvelous passage on another Taoist, Shen Tao (who, as the Americans would say, was even more "way out"). The translation is that of Arthur Waley.[1]

[1]*Three Ways of Thought in Ancient China. London:* George Allen & Unwin Ltd., 1939.

Shen Tao, discarding knowledge and the cultivation of self, merely followed the line of least resistance. He made an absolute indifference to outside things his sole way and principle. He said, "Wisdom consists in not knowing; he who thinks that by widening his knowledge he is getting nearer to wisdom is merely destroying wisdom."

His views were so warped and peculiar that it was impossible to make use of him; yet he laughed at the world for honoring men of capacity. He was so lax and uncontrolled that one may say he had no principles at all; yet he railed at the world for making much of the Sages. He let himself be pounded and battered, scraped and broken, be rolled like a ball wherever things carried him. He had no use for "Is" and "Is not," but was bent only on getting through somehow. He did not school himself by knowledge or thought, and had no understanding of what should come first and what last, but remained in utter indifference. Wherever he was pushed he went, wherever he was dragged he came, unstable as a feather that whirls at every passing breath of the wind, or a polished stone that slides at a mere touch.

Yet he remained whole; nothing went amiss with him. Whether he moved or stood still, nothing went wrong, and never at any time did he give offense. What was the reason of this? I will tell you. Inanimate objects never make trouble for themselves. They do not burden themselves with knowledge, and yet never whether in motion or at rest do they depart from what is reasonable, and for this reason they never go wrong. That is what he meant by saying "all that is necessary is to make yourself like an inanimate object; do not try to be better or wiser than other people. A clod of earth cannot lose its way."

The great men of the day used to laugh at him, saying that Shen Tao's principles were better suited to the dead than to the living, and might astonish, but certainly could not convince. (Chuangtse, XXXIII)

The above is one of my favorite passages. It is interesting how differently people react to it! One person to whom I read it (a teenager whose parents are overly anxious that he do well in school and have been pestering the poor life out of him!) found this passage delightful and consoling. He said: "The passage is great! That's just how I feel; the guy reminds me exactly of me!" But other people are horrified! One friend of mine, who is successfully established in a profession, did not like the passage at all. He said, "I find it quite undignified to liken a human being to a clod of earth." I must say, this reaction saddened me! What could possibly be *undignified* about a clod of earth? For that matter, what could be *dignified* about a clod of earth? If anything *transcends* these silly categories of dignity and nondignity, it is surely a clod of earth!

Let me digress a moment on the subject of "dignity." I find nothing more undignified than a person who consciously values "being dignified." Does the Tao value dignity? Of course not; it is far too dignified—in a purely natural and spontaneous sense—to have any such undignified notion as "dignity." There is perhaps such a thing as "true dignity" and "false dignity," or perhaps "natural dignity" and "artificial dignity." As perfect examples of natural dignity, I would cite my dogs Peekaboo, Peekatoo, and Trixie. Everyone who comes to the house comments on them. As one dear friend said, "They must be seen to be believed." Another said about Peekaboo (the mother of the others) "She is always a lady." Now, Peekaboo was not *trained* to be a lady! She has had true, natural dignity ever since she was a little puppy. But she, like the Tao, does not place a *premium* on dignity. She seems totally unconscious of the "concept" of dignity. She would not regard it as "beneath her dignity" to be undignified, and in this lies her true dignity. And so it is with humans. When my wife and I visited Greece, one of the things we found most outstanding about the people of all strata of society was their beautiful natural dignity—totally without any affectation. It is the element of affectation, as well perhaps as elements of snobbishness and "false pride," which characterize artificial or false dignity as opposed to natural or true dignity. Natural dignity comes close, I would say, to what the Christians would call "humility." And I don't believe that a person with "true" dignity would regard it beneath his dignity to be a clod of earth.

After this digression, I would like to return to Chuangtse's passage on Shen Tao. Those of you who find the passage disturbing, depressing, or shocking rather than enlightening or inspiring, would you like to know something of my reasons for finding it otherwise? Well, let us take some of the parts line by line: Consider the first line: "Shen Tao, discarding knowledge and the cultivation of self, merely followed the line of least resistance." Now, I love the idea of following the line of least resistance! Here I identify myself completely with Shen Tao, for all my life I have followed the line of least resistance, or at least *felt* as if I did. Now, I am well aware of the Puritanical attitude toward the notion of following the line of "least resistance"; this, they say, is the path to indolence, inactivity, alcoholism, drug addiction, sexual debauchery, and ultimate damnation. The Taoist attitude is totally different! To the Taoist, "resistance" means resistance to the Tao—or better still, "friction" with the Tao. The Taoist ideal is to avoid friction with the Tao, and it is only erroneously thought by those who do not understand, that this means "passivity" rather than "activity." But actually it means neither. The point is that if the Tao happens to be at rest at the moment, and you persist in actively moving, then you will be moving through it, hence generating friction with it. On the other hand, when the Tao itself is in motion, and you are passively at rest, then the Tao will be brushing past you, so again there will be friction. The idea, then, is to guide your movements by the movements of the Tao; in this way you will generate no friction but rather will be in tune with or in harmony with the Tao, as well as avail yourself of the tremendous power that the Tao can provide.

This same idea is nicely expressed in the I Ching (§52):

"True quiet means keeping still when the time has come to keep still, and going forward when the time has come to go forward. In this way rest and movement are in agreement with the demands of the time, and thus there is light in life."

Next consider the line: "Wisdom consists in not knowing; he who thinks that by widening his knowledge he is getting nearer to wisdom is merely destroying wisdom." Now, it might seem strange that one who places a high premium on learning and cultivation of the mind should find this line so attractive. Superficially, it might well appear to be anti-intellectual. But it does not have to be interpreted this way. The real point is that much knowledge of facts does not necessarily constitute true wisdom. Indeed, if a person cultivates knowledge for the *purpose* of becoming wise, he is extremely unlikely to succeed. Indeed, any knowledge cultivated for any purpose whatsoever is unlikely to be of much benefit. Another even more important point is that knowledge, despite its obvious benefits, does sometimes have the unfortunate effect of only removing us from that which is of primal importance. This idea is beautifully expressed elsewhere in the following passage of Chuangtse.[1]

> The knowledge of the ancients was perfect. How perfect? I will tell you. At first they did not know that there were things. This is perfect knowledge to which nothing can be added. Next they knew that there were things, but did not yet make distinctions between them. Next they made distinctions between them, but did not yet pass judgements upon them. When judgement was passed, Tao was destroyed. With the destruction of Tao, individual preferences came into being.

Next consider the line: "His views were so warped and peculiar that it was impossible to make use of him." I haven't much to say about this, other than that I love the style in which it was written! The idea of "making use" of somebody!

[1]Composite translation.

Now consider the line: "He was so lax and uncontrolled that one may say he had no principles at all." Now, to the more Puritanical mind, absence of "principles" is associated with unprincipled behavior. But to the Taoist, this is not so. A typical Taoist remark such as "Give up advertising goodness and duty, and people will regain love of their fellows" well illustrates this point. The whole idea is that often the very holding of a "principle" may be the major cause of acting out of accord with the principle. To give some examples: A person whose principle it is to be "spontaneous" will hardly succeed by "trying" to be spontaneous. A person who believes that he "should" love his neighbor as himself will probably never succeed in doing so; he might succeed in being just to his neighbor, but will hardly succeed in really loving him—the reason being that the motivating force behind his "trying" to love his neighbor is not love of his neighbor but adherence to some moral principle. Now, if he would stop "trying" to love his neighbor, if he would not have it as a "principle" that he "should" love his neighbor, if he would stop trying to change himself and stop trying for moral improvement, in short if he would let himself alone, then there is a very good chance that he would gradually grow more loving as the years went on. Thus I say that having "principles" often defeats the very purpose of the principles. So therefore when I read that Shen Tao was "unprincipled," I interpret it not as unprincipled behavior, but as simply not behaving this way "on principle." For example, he may have been extremely truthful, and yet truthfulness may not have been one of his "principles." Or he may have been inoffensive, but not necessarily "on principle." Indeed, we read later that "never at any time did he give offense," so I must be right! So, I hope I have adequately explained the difference between having a principle and acting according to the principle. If you want a still more drastic example, a clod of earth is inactive, but not on "principle": it does not "will" to be inactive (just as the sage is quiet because he is not moved, not because he wills to be quiet). But, you will shout, "a human being is *not* like a clod of earth; the comparison is most undignified!" Well, shall we argue about that again? I say a human being *is* like a clod of earth—at the least the best of them like Shen Tao are! Yes, I believe *on principle* that one should be exactly like a clod of earth. And it is because I have this "principle" that I fail to be the clod of earth I so love!

I am getting weary of this "line by line" analysis and shall content myself with a few general remarks. I found the bit about letting himself "be pounded and battered, scraped and broken, be rolled like a ball wherever things carried him"—I found this bit screamingly funny. Some will find it more pathetic than funny, but I wish to go on record as declaring it funny. Some will claim he was "masochistic," but I do not believe it! I would say rather that he had the good sense to cope with a few scrapes in life and did not find it beneath his "dignity" to be rolled about like a ball. His scrapes couldn't have been too serious, for it was said that through it all he remained whole. And anybody who can remain whole through it all has my respect!

The most profound line of the whole passage is "He had no use for 'Is' and 'Is not,' but was bent only on getting through somehow." This line is so profound, that I can find no words adequate to discuss it.

A Blade of Grass Does Not Quiver

☐ *A blade of grass does not quiver*
Unless there is a wind,
Or someone pushes it,
Or unless there is another reason.
Rare is the blade
Which quivers for no reason whatsoever.
These blades are my favorite!
They quiver only because they want to.

The Sage Needs Nothing

☐ Aristotle in his Ethics makes the following superb remarks: "The man in contemplation is a free man. He needs nothing. Therefore nothing determines or distorts his thought. He does whatever he loves to do, and what he does is done for its own sake."

All I can say to these words is: "Ah! Yum-yum! How perfect! How is it that Aristotle in four short sentences has put my entire philosophy of life into a nutshell?"

A skeptic, at this point, might well reply: "Fine! It sounds very nice, and all that, but is it really *true*? A man in contemplation needs *nothing*? You mean he does not even need food? He can live without food?"

Rather than give a "rational" answer to this skeptic, I would prefer to quote a Chinese philosopher who, I think, went even one better than Aristotle. He says: "The Sage needs nothing, not even life."

I like this latter saying even better! It puts my entire philosophy of life into an even smaller nutshell. I was recently discussing all this with a sociologist. Now, this sociologist is of the so-called "rational" type, and so he, of course, took exception to this. In particular, he could not understand what could be meant by the sage's not "needing" life? He asked, "What on earth does it mean to '*need*' life?" Then he thought for a while, and interpreted it to mean being "indifferent" to life and death. To me, this interpretation seems all wrong! If the Chinese philosopher had instead said, "The Sage is indifferent to life and death," I would not have liked it nearly as well! Well then, since I have rejected this interpretation, what alternative interpretation do I have to offer? Curiously enough, I have no alternative explanation to suggest. I can give no "rational" explanation of what it means to "need" life, yet I understand what it means perfectly! There are those, of course, who would sharply deny this. They would say that unless a person can "explain" what he means, then he really does not mean anything at all. (This is like the so-called "Socratic fallacy" which is that unless one can define a term, one does not mean anything by the term.) My point of view is about the opposite. I believe there are many highly significant terms and sentences such that any attempt to explain them only obscures their true meaning.

I find it of interest that many people who vehemently deny that a certain word is meaningful will themselves use the very word in a psychologically unguarded moment. The following brief dialogue, though imaginary, is not too far removed in spirit from dialogues I have actually heard.

Rationalist: I don't understand you! I really don't understand you at all! You keep on using this word "mystical" but have never once defined what it means. Frankly, I have no idea what it means, nor what it even could mean. Indeed, I suggest that the word "mystical" has no meaning whatsoever.

Mystic: I think you are looking far too far afield for its meaning. Actually its meaning is not the least bit subtle, but is as plain as daylight. It lies under your very nose, and for this reason you overlook it.

Rationalist: I still can't understand you! You are being too mystical.

Why Do You Wish to Be Enlightened?

☐ Suppose you wish to get "enlightened." You do it, e.g., through the path of Taoism, Zen Buddhism, or some form of yoga. Now, suppose a person asks you: "*Why* do you wish to get enlightened? What is your *purpose* in getting enlightened? For whose sake are you seeking enlightenment? Is it to benefit *you* or to benefit society?" Now, the amazing thing is that however you answer, you will find yourself trapped! Suppose you answer, "For *my* sake." Then people will descend upon you with the fury of hell and say: "You selfish egotist! Just as I thought! You care only about yourself! You don't give a damn about others! So *you* with your enlightenment will become internally peaceful and serene, but how will that help the problems of the world? So, I was right about the whole subject of enlightenment! It is a purely *selfish* enterprise!"

On the other hand, suppose you answer, "I primarily wish to help others, but I must first get enlightened myself before I can spread enlightenment to others." Well, if you give *that* answer, then people will descend upon you with the fury of hell and say: "You arrogant, conceited egotist! Just as I thought! So it is up to *you* to enlighten others, eh? *You* the great prophet must go forth to teach others! You cannot be content to live a simple life of your own—ah no! You have to be in the limelight; you have to be this great 'saint' and 'sage' and be the 'savior' of mankind! So I was right! This whole 'enlightenment' business is just to feed people's vanities!"

What can you do when people speak to you like this? Alas, I wish I could tell you; I wish I could help you! But unfortunately I am not yet a complete sage.

Satori

☐ In Eduard von Hartmann's magnificent book *Philosophy of the Unconscious* [1] (which, incidentally, all of you should read) occurs the following wonderful passage:

> To wit, *conscious thought* can comprehend the identity of the individual with the Absolute by a rational method, as we too have found ourselves on the way to this goal in our inquiry; but the Ego and the Absolute and their identity

[1] London: Kegan Paul, Trench, Trubner & Co. Ltd., New York: Harcourt, Brace & Company, 1931.

stand before it as three *abstractions* whose *union* in the *judgment* is made probable, it is true, through the preceding proofs, yet an *immediate feeling of this identity* is not attained by it. The *authoritative belief* in an external revelation may credulously repeat the dogma of such a unity—the living feeling of the same cannot be engrafted or thrust on the mind from without, it can only spring up in the mind of the believer himself; in a word, it is to be attained neither by philosophy nor external revelation, but only mystically, by one with equal mystical proclivities, the more easily, indeed, the more perfect and pure are the philosophical notions or religious ideas already possessed. Therefore this feeling is the concept of mysticism.

The whole content of this passage was alternatively expressed by a Zen master when he once said about Satori (or enlightenment): "It is like drinking water and knowing *yourself* that it is cold."

How true! The enlightened man does not give a logical *proof* that the water is cold, nor does he perform elaborate scientific experiments to prove that the water is cold with probability such and such, and he certainly doesn't have to take it on "faith" or "authority" that the water is cold; he directly *knows* that the water is cold. Nor does he require a prior precise definition of "cold," nor have to answer such questions as, "At precisely what temperature should water be declared cold?" No, he doesn't have to do any of these things to know that the water is cold. Is this not remarkable?

On Trying to Obtain Enlightenment

☐ There was a delightful scene in the movie version of *I Love You, Alice B. Toklas*, in which the main character—acted by Peter Sellers—was walking on the beach with his guru, and the guru said, "It is very important that you *find* yourself." The man replied, "I am *trying* to find myself." The guru replied, "Only after you have stopped trying will you have found yourself." The man then replied in a voice of pleading desperation, "I am *trying* to stop trying!'

Had I been the guru, I would have added, "Stop trying that, too!" I don't think it very possible that he could have tried *that*. This is, I don't see how one could very easily try to stop trying to stop trying, since things at this point are getting conceptually too complicated.

This calls to mind the following story: A Zen student went to his master in a very exuberant state and said: "Master, it is wonderful! I have gotten rid of all conceptualization, all discursive thought; I have gotten rid of everything! What do you think of *that*!" The master replied, "Get rid of that, too!" The student replied: "But Master, you don't understand! I have gotten rid of *everything!*" The master replied, "Very well, since you won't get rid of that, by all means carry it with you."

The futility of "trying" to get enlightened is best expressed by the following passage, a portion of a discourse of the Chinese Zen Master Tsung Kao:

"If you use one iota of strength to make the slightest effort to obtain Enlightenment, you will never get it. If you make such an effort, you are trying to grasp space with your hands, which is useless and a waste of your time!"

Why Did You Go to the Zen Monastery?

☐ It should be clearly recognized that despite many divergences and oppositions, Mysticism and Logical Positivism are closely allied. Both, for example, repudiate, or seek to eliminate, Metaphysics. For example, the Buddha, when asked whether the soul survives bodily death, categorically declared such metaphysical questions should not even be asked—they have no bearing on the problem of deliverance from pain (salvation). Many questions—such as the mind-body problem, free will versus determinism, existence and survival of the soul—are rejected by positivist and mystic alike, the former on the grounds of lack of cognitive meaning, and the latter on somewhat divergent grounds, one of them being that these questions are too "dualistic." Even the very question of monism versus dualism might be rejected by a mystic on the (paradoxical?) grounds of being too dualistic!

Yet, of course, the positivist will reject many of the assertions of some mystics as being totally meaningless (e.g., "time is unreal," "all is one," "there is a cosmic consciousness"). So, of course, there are also major divergences between the two attitudes. But I am here concerned more with their similarities than divergences.

There is an even weirder relation which I suspect exists between the two, which sort of combines opposition with agreement, and is each of them precisely by virtue of the fact that it is the other! Mystical as this last statement may sound, I think I can illustrate my feeling behind it by the following imaginary dialogue.

Positivist: How good to see you again! Twelve years! And *all* this time you really spent in a Zen monastery?
Mystic: Oh yes!
Positivist: How come you did it?
Mystic: What do you mean?
Positivist: Don't you recall the long protracted conversations we used to have about mysticism, and the fact that after much resistance you finally acknowledged that it is all utter nonsense?
Mystic: Of course I remember!

Positivist: Then how come you went anyhow? What were your reasons?

Mystic: Reasons? I don't know that I had any reasons, I just went!

Positivist: Well, did you accomplish your goal? Did you attain this thing called "enlightenment" or "Satori"?

Mystic: Yes.

Positivist: How do you know you attained it? What proof do you have?

Mystic: Proof? None that I know of. The fact that a certified Zen master attested to my Satori is hardly *proof* that I attained Satori. How do I *know* I attained Satori? I don't know how I know, I just know. Satori is one of those things which when once attained requires absolutely no proof for its recognition.

Positivist: I have heard much about the so-called ineffability of Satori. But is there really *nothing* you can say about it to give me at least *some* idea of what it feels like? Did it not give you *any* insights which you can in some way describe?

Mystic: Oh certainly! I don't think Satori is *all* that ineffable! Much of it—perhaps most of it—is, but certain aspects are indeed describable. And one important insight in particular, perhaps the most important insight of all, certainly is describable and will be of particular interest to you. Indeed, this is the main purpose of my visit.

Positivist: I am all ears!

Mystic: Well, as you said, twelve years ago we spent many and many an hour arguing the validity and meaningfulness of various mystic claims. You set up many formalized languages, and we together made many charts of the various possible meanings of some of the ambiguous terms used by mystics, and you carefully demonstrated that under all the possible combinations of meanings, none of the mystic assertions made any sense. And as you said, I resisted all this month after month using counter-argument after counter-argument. But you demolished all my counter-arguments one after the other until finally I had to admit complete and total defeat. Not only did you set up a precise definition of meaning and show that mystic statements are meaningless according to this definition, but you were even sufficiently charitable to invite me to propose my own definition of "meaning." I proposed many, many of them, and you then patiently went through the time-consuming process of showing me that according to every one of these definitions of "meaning," all the mystical assertions in question are meaningless—and this, mind you, under every combination of meanings of all the ambiguous terms of the statements. Thus your analysis was absolutely exhaustive (and, I might also add, *exhausting!*). So finally reason compelled me to admit that I was wrong and you were right. But what happened was this. Although on purely rational grounds I was forced to admit that you were right, on purely emotional or intuitive grounds I *felt* with every ounce of my being that you were wrong. Although I could find not the slightest error in your argument, I could not help but feel that somewhere there lurked a serious error which I was simply not intelligent enough to uncover, or that your whole basic approach was somehow wrong. But my reason told me that this simply could not be; to claim it was would literally involve a contradiction. So which should I trust, my reason or my intuition? This situation plagued me for months, and got more and more unbearable. Finally I could stand it no longer, and I decided to trust my intuition, and so I entered a Zen monastery.

Positivist: Oh, so you *did* have a reason for going! Why did you tell me before that you went without a reason?

Mystic: I did not say that I went without any reason, I said that I did not know of any *reasons* why I went. Indeed I had no *reasons*; my *reason* told me *not* to go. I went following my *intuition*, not my *reason*.

Positivist: How unreasonable of you!

Mystic: Obviously! On the other hand, if I had not gone, it could be said "How un-intuitive of you!"

Positivist: Well, at any rate, you have not yet told me of what great insight you attained upon achievement of Satori.

Mystic: I am coming to that! I went to the Rishi and after much discussion of my problem, he decided it best that I pick my own koan.[1] Naturally I picked the question, "Does mysticism have any meaningful validity?" And so I struggled with this question for twelve hard years, not from the *logical* point of view which I had already settled with you, but from the point of view of understanding my *real* feelings concerning the matter. At last I attained Satori, and answered the koan. The answer was "No, mysticism does *not* have any meaningful validity!" I rushed excitedly to my master exclaiming, "So mysticism *is* a lot of nonsense! It *is* a lot of nonsense!" And my master embraced me with tears of joy, exclaiming, "At last you are a true Mystic!" Then I reflected on all the past conversations I had with you. So you *were* right after all!

Positivist: Then I don't see what you have gained. You already admitted I was right twelve years ago!

Mystic: Of course I *admitted* you were right! My *reason* told me that you had to be. But as I told you again and again, my *intuition* was totally incapable of believing that you were right. Now, for the first time in my life, I am capable of *intuitively* knowing that you are right.

Positivist (stunned): I hardly know what to say.

Mystic: What would you like to say?

Positivist: I mean that I do not know whether to regard your present situation as a triumph for me or as a defeat. On the one hand, you now realize *fully* that I was right—this, I guess, is a triumph. On the other hand, the reasons I gave you twelve years ago were evidently not good enough for you. No, you had to acquire your present mystic intuitive insight into the meaninglessness of mysticism by employing the very mystic techniques I so deplore. So from this point of view, the situation is for me a defeat. On the *verbal* level, you acknowledge the correctness of my position, yet your very method of

[1] A koan is a problem given by Zen-Masters which has no logical solution; its purpose is to force the realization of the futility of logic in dealing with ultimate reality.

arriving at this conviction can be interpreted as an indication that in some sense I must be wrong. So therefore, as I said, I do not know whether I have triumphed or have been defeated.

Mystic: I am sorry that you structure the situation in such a personal and competitive manner. One could also ask whether *I* have been defeated or have triumphed. If the recognition that my former opponent is right is to be looked upon as a defeat, then I have been defeated. But on the other hand, the very fact that mystic techniques have succeeded in bringing my intuition into harmony with my reason, whereas rational techniques have failed to do this, does show a sort of triumph of mysticism over rationalism. So I too can be said to have triumphed in one sense and been defeated in another. But truthfully I feel neither a triumph nor a defeat. In a way *both* of us can simultaneously claim triumph and defeat—indeed, it might be said that each of us has been defeated by virtue of his very triumph, or, to put it more optimistically, we have each triumphed by virtue of our very defeat. But it is better still not to even use the categories of triumph and defeat in the present context. Something far more important has emerged from all this than the purely personal questions of whether you or I have won or lost. What has truly emerged is this.

Mysticism is *not* a set of beliefs, but purely a state of being, like being short or being fat, or being musical or being humorous. There is nothing true or false about *being* in a given state; it is only *assertions about* the state which are true or false or sometimes meaningless. So the important thing about the mystic is not his assertions *about* the state, but his being *in* the state. So of course most of the mystic *assertions* about the state are both positivistically meaningless and of no real importance to the *true* mystic, since they simply fail to hit the mark. I spent so much time with you discussing only statements *about* the mystic state instead of actually entering the state and *knowing* what it is really like.

Positivist: So now you have entered the state and really know firsthand what it is really like.

Mystic: Oh yes!

Positivist: And you cannot tell me more about it?

Mystic: I can and will in due time. But for the present, the part which I believe is most helpful for you to know is that entering the state implies a *full* realization of the inadequacy of sentences to describe the state. All those sentences are totally meaningless, just as you have always said!

VII THE FRUIT OF KNOWLEDGE

The Fruit of Knowledge

☐ The reason Adam ate of the fruit of knowledge was that he didn't *know* any better. Had he had just a *little* more knowledge, he would have known enough not to do such a damn fool thing!

Can we return to the Garden of Eden? Well, if we returned *completely*, if we entered again into the complete state of innocence, we would no longer have the knowledge to prevent us from eating the apple again. And so again we would fall out of grace. It seems, therefore, that to regard the Garden of Eden, the state of innocence, as the *perfect* state is simply a mistake. It has the obvious imperfection of being internally unstable and self-annihilating.

Too bad there weren't two trees of knowledge in the Garden of Eden, a big tree and a little tree. The only knowledge to be imparted by the little tree should be, "It is a mistake to eat of the big tree."

What Is There?

☐ Someone once had the following idea for a question on a physics examination: "Define the universe, and give two examples."

I thought of the following variant of this question: "Define an *entity* and give a counterexample."

Quine starts his famous essay "On What There Is" with the words: "A curious thing about the ontological problem is its simplicity. It can be put in three Anglo-Saxon monosyllables: 'What is there?' It can be answered, moreover, in a word—'Everything.'"

This reminds me very much of an incident in Oscar Mandel's delightful book, *Chi Po and the Sorcerer*.[1] (One thing

[1]Charles E. Tuttle, Co., Rutland, Vermont, and Tokyo, Japan: 1964.

that attracted me to this book is the subtitle: "A Chinese Tale for Children and Philosophers." I bought this book because I felt qualified as a reader on both counts—certainly on the first, and possibly on the second.) In this story, the boy Chi Po is taking painting lessons from the sorcerer Bu Fu. At one point, Bu Fu is looking at Chi Po's painting and says: "No, no! You have merely painted what *is*. Anybody can paint what is! The real secret is to paint what isn't." Upon which Chi Po was very puzzled and said, "But what is there that isn't?"

Is Man a Machine?

☐ Recently I was with a group of mathematicians and philosophers. One philosopher asked me whether I believed man was a machine. I replied, "Do you really think it makes any difference?" He most earnestly replied, "Of course! To me it is the most important question in philosophy."

I had the following afterthoughts: I imagine that if my friend finally came to the conclusion that he *were* a machine, he would be infinitely crestfallen. I think he would think: "My God! How horrible! I am *only* a machine!" But if *I* should find out I were a machine, my attitude would be totally different. I would say: "How amazing! I never before realized that machines could be so marvelous!"

Intuition versus Reason

☐ I find it remarkable that people argue about this! Argument involves *reason* which is already loading the dice. When reason itself is on trial, one can hardly expect reason to be the judge! When people on the side of reason claim reason to be more reliable than intuition, they give *reasons* to support their belief. Those on the side of intuition claim their intuition tells them that intuition is superior to reason.

Can reason ever be in conflict with intuition? Why certainly! There are false reasons and false intuitions. But valid reason obviously cannot be in conflict with valid intuition, since truth cannot be in conflict with truth.[1] The real question for any person is which is more reliable—his intuition or his reason! For another person to say "you should trust your intuition" or "you should trust your reason" is obviously foolish; how does *he* know?

Well, which *should* the person trust? How do I know? Well, how should *you* decide which to trust? By consulting your reason or your intuition? I don't know that either.

One thing I do know: Certain people called "rationalists" make the definite claim that the only reliable road to knowledge is through science and reason. This claim is one of the most remarkable dogmas I have ever heard! I have seen many *reasons* given to support this claim, but they are unbelievably bad! Yet, of course, the claim *may* be true. It may be true but not provable (not even with significantly high probability). I myself do not know of a single valid reason to support the claim. And my intuition tells me that it is exceedingly unlikely. But my intuition may be wrong. In which case, the claim is true, despite it unprovability by its adherents. I would not be surprised if the rationalists recognize the truth of this claim by virtue of some valid mystical intuition which I lack.

Spontaneous Hedonism

☐ My introduction to philosophy—or rather one of my introductions to philosophy—was somewhat unfortunate. I was about fourteen years old at the time. A philosopher several

[1] Or can it?

years my senior asked me, "Raymond, what do you want most out of life?" I instantly, spontaneously, and unhesitatingly replied, "Pleasure!" The philosopher then proceeded to try to show me that this answer was "irrational." He said: "Suppose you had before you a glass of very delicious poison. You know that it is fatally poisonous but that it would afford you great pleasure to drink it. Would you?" I replied, "Of course not!" He said, "Why?" I said, "Obviously, because I know that the stuff would kill me, and hence that whatever pleasure I would lose by not drinking it would be negligible compared with all the future pleasures in life I would miss by dying prematurely." He said: "Ah, so it is not pleasure, but *reason* which is the criterion of your actions!" Now it so happened that I did not know the meaning of the word "criterion" at the time, but was ashamed to admit it. So I was therefore nonplussed and could offer no reply. The philosopher then triumphantly continued, "So you see, you are really living for *reason*, not for pleasure!" A week later our conversation continued, and I told him that I did not know the word "criterion." After he explained it to me, I at once saw the idiotic fallacy of his whole argument, and I must have been furious at the deception! I exclaimed: "Oh, for heavens sakes! Your argument does not show that I am living *for* reason, or that reason is the criterion of my actions! No, no; pleasure is indeed the criterion. It is just that I *use* reason to help me obtain the greatest amount of pleasure. Thus reason is the tool, not the end. To put it otherwise, I might live *by* reason, but not *for* reason." He could find no reply to this, and sheepishly, yet with a strange embarrassed but disdainful smile, said "Oh!"

This memory has left a very bad taste in my mouth for philosophy! What was wrong with this philosopher? Why couldn't he distinguish between *using* reason in one's life and living *for* reason as the aim of one's life? Was he being just dense, or was he being sophistical? I still don't know! At any rate, the conversation then developed along more helpful lines. He told me that I am what is known as a "hedonist," and he then carefully distinguished between the doctrine of "ethical hedonism," which asserts that one *should* have pleasure as the goal of one's life, and the doctrine of "psychological hedonism," which claims that the pursuit of pleasure *is* the goal of our lives, whether we realize it or not. According to this doctrine, even our so-called altruistic acts are in reality purely selfish, because they are undertaken only for the pleasure it gives us to act altruistically.

When I first heard about *ethical* hedonism, I burst out laughing! The idea that one *should* do the very thing one wants to do struck me as so funny! Psychological hedonism struck me much more favorably, and this is the ethical philosophy I tended to hold most of my life. I no longer do. Let me try to give you some indication why.

In the first place, I am extremely skeptical of our ability to know the motives of our acts; indeed I am rather of the opinion that the vast majority of our more important acts are performed *with no motive at all*! I know this sounds strange to most Westerners—particularly those who are psychoanalytically oriented—but this is nevertheless the way I see it. Certain of my acts have a "purpose," but the more important ones do not. For example, if I am walking to the library to get a book and someone asks me why I am walking, I can honestly say "for the purpose of getting such and such book." But if someone asks me what is my purpose in *reading* the book, I regard the question as both hostile and ridiculous. I wish to go on record as saying that in general, I do not read books for any *purpose*. Yes, it has sometimes happened that when I had to read a book in order to pass some stupid exam, then I had a purpose in reading the book—viz. to pass the exam. But in general when I read a book which I love and enjoy, I think it is completely misleading to say that I am reading it for some "purpose." The psychologist may say that I really do have a purpose but that for some reason my knowledge of the purpose has been repressed, so the purpose is purely "unconscious." But why must I listen to such psychologists? And the psychological hedonist will tell me that when I read a book which I enjoy, and claim to have no purpose in reading it, my *real* purpose is the enjoyment which the book gives me. But why must I listen to the psychological hedonist? Just because I do get pleasure from reading the book and have no other motive, why does it follow that I have *this* one, i.e., the pleasure of reading? If the hedonist answers, "But if it were not for the pleasure of reading the book, what other reason would you have for reading it?" then I would say that he is totally begging the question by assuming that there must be *some* reason. Why does everything have to have a reason? I realize only too well that the whole Western so-called "rational and scientific tradition" has it that everything which happens has some cause or reason, if only we can find it. It is sadder still that most Westerners *delight* in this fact! Well, my attitude is much more like that of many of the classical Chinese poets who regard life—even human life—as being like floating clouds—sometimes assuming this form and sometimes that, or sometimes no form at all. However, I digress (just like a floating cloud).

How does the psychological hedonist *know* that whatever I do is for the *purpose* of obtaining pleasure? Is this assertion supposed to be analytic or empirical? Of course, one can easily make it analytic by giving a purely operational or behavioristic definition of "pleasure": simply define an act to be "pleasurable" if it is the type of act which the organism tends to pursue. Then, of course, psychological hedonism becomes true but at the cost of total triviality. No surely the psychological hedonist means more than this when he asserts that pleasure is always the goal of our actions. But then, on what empirical grounds does he know this? What experiment could possibly be devised which could ascertain whether pleasure is always the goal of our acts? Now, I know by direct introspection that pleasure is *sometimes* the goal of my acts. But to conclude that pleasure is *always* the goal of my acts strikes me as completely unwarranted.

There is a variant of psychological hedonism, which we might call "instantaneous hedonism," which strikes me as far more plausible. And that is the doctrine that we always do precisely that which gives us the greatest pleasure *at the very moment!* But this is very different from saying that pleasure is our *goal.* It seems to me that the goal of an act must always occur later in time than the act itself. If, for example, I am going out to dinner and I choose one restaurant rather than another, it is likely that I do so with the *goal* of having the more pleasant meal. But while I am actually eating the meal, it strikes me as sort of odd to say that I do so for the *purpose* of the pleasure it gives me. Yet, it is probably the most pleasurable thing I could do at the moment. Thus instantaneous hedonism seems more plausible than the more usual psychological hedonism. Actually, the epistemological status of instantaneous hedonism is a bit strange. Is it a tautology or is it an empirical statement? I really don't know. Yet it has, at least for me, a sort of weird ring of truth to it. I would not go so far as to say that I necessarily *believe* it. But I would not be surprised if—unlike psychological hedonism—it were true.

Incidentally, I do not believe, as do some, that psychological hedonism is something inconsistent or "demonstrably false." I feel virtually certain that it is not disprovable. The reason that I do not believe it is simply that there is no good reason why I should believe it—and besides, it strikes me as rather counterintuitive.

There are some who object to hedonism on ethical grounds. They feel that hedonism is somehow or other an apology for "selfishness." To me, this charge is totally unfair! I judge people by their actions, and the avowed hedonists I have known are hardly any more selfish than their opponents. If anything, they are only more honest about the question. Of course *some* hedonists are what I would call 'intellectually dishonest" (or perhaps they are just plain stupid) when they assert that all people are equally selfish since all they do is to pursue their own pleasures even when they are acting in an allegedly "altruistic" manner. Surely, now, this sophomoric fallacy is obvious to almost all of you! Even if psychological hedonism were absolutely true, it would not in the least follow that altruism is but another form of egoism! An altruistic act (as I understand it) is by definition one which is performed with the intention of benefiting others. Even if it were true that one wishes to benefit others only because of the pleasure to oneself in doing so, this nowise obviates the distinction between an intention to benefit others and an intention which is not an intention to benefit others. To say they are *really* the same (since they are both ultimately motivated by the desire for pleasure) is just as silly as saying that red and green are *really* the same, since in the last analysis they are both colors. Thus it is ridiculous to use hedonism to try to prove that there is no such thing as altruism. But why am I saying all these obvious and elementary things?

There is another form of hedonism I wish to consider, which I would like to call "natural hedonism." This "natural" hedonism is not in any sense a doctrine, but is rather an attitude toward life. Unlike psychological hedonism, it does not say how we do behave, and unlike ethical hedonism, it does not prescribe how we should behave. Since it is not really doctrine, I should perhaps speak of the "natural hedonist" rather than of natural hedonism. Well, to me, the natural hedonist is the one who says "What I value most in life is pleasure." There is obviously nothing to prove or refute about this; it is simply a statement of fact. Some people do value pleasure more than anything else; others do not—it is as simple as that. Of course, we might classify natural hedonists into two types: the natural individual hedonist, who values his own pleasure above everything else, and the natural social hedonist, whose main value is people's pleasure (or better still, the pleasure of all sentient beings).

I must again emphasize that natural hedonism is not something which can be either proved or refuted; it can only be sanctioned or frowned upon. I guess relatively few moralists would condemn natural social hedonism, whereas I imagine most moralists would sharply condemn natural individual hedonism. But is natural individual hedonism necessarily all that bad? What about a person who happens to be extremely sympathetic and whose main pleasure in life is to make others happy? Surely you would not condemn him for pursuing *this* pleasure! Would it make any practical difference if such a person claimed to be an individual hedonist rather than a social hedonist? That is to say, does it make much difference for a person who does in fact value other people's happiness, whether he does so for the pleasure it gives him or whether he values other people's happiness as an end in itself?

To summarize my main points: (1) ethical hedonism is simply silly!; (2) psychological hedonism is an unwarranted dogma, but it would be extremely difficult, if not impossible, to disprove it; (3) instantaneous hedonism is more plausible, but it would still be very difficult to prove or disprove; (4) natural hedonism (which I feel is the most honest form of all) is not really a doctrine, and perhaps for this reason, should be taken the most seriously; (5) no form of hedonism is really "immoral," and it is not very nice of moralists to say that it is.

Why Argue about a Definition?

☐ I believe it is sometimes very important that one *should* argue about a definition. This is far from a merely pedantic enterprise. The words we choose for our own use have a major effect in coloring our thinking and in determining our emotional attitudes toward things. Let us consider the following dialogue.

Naturalist: To me, nature is everything. We are immersed in the idea of nature. Everything we do is part of nature. Nature

127

flows right through us. We *are* Nature, or at least of nature. Nature is everything.

Theist: I disagree. To me, God is the ultimate reality. Nature is the physical; God is the spiritual.

Naturalist: I do not deny the spiritual reality which you affirm. If it exists, then I would call it part of nature. To me, Nature is *everything*.

Theist: Then you are not using the word "nature" in its proper sense. God *created* Nature, but God is not the *same thing* as Nature.

Naturalist: What do you mean "proper sense"? Some indeed use the word "nature" as you do, others use it as I do. I doubt that the meanings of either of the words "God" or "Nature" have been sufficiently standardized to justify reference to "proper" or "improper" usage.

Theist (a bit exasperated): But "God" and "Nature" simply mean different things.

Naturalist: I do not deny that what *you* mean by "God" does indeed differ from what *you* mean by "Nature." But I'm not so sure that what you mean by "God" differs from what *I* mean by "Nature."

Theist: Then why should we argue over a mere definition? We both accept the existence of the ultimate reality—whatever that may be. Why not call it "God"?

Naturalist: Why not call it Nature?

Theist: (After a pause) No, that would never do. After all, when I pray, it would sound sort of silly to begin with "Dear Nature" instead of "Dear God."

Naturalist: And for me, it would never do to use the word "God."

Theist: Why not?

Naturalist (after a pause): I have always called myself an atheist. And I think it would sound strange for an atheist to say, "I believe in God."

Theist: But you *do* believe in the ultimate reality. What's the difference what *name* you use for it? Why not call it "God"?

Naturalist: Of course I believe in the ultimate reality; who doesn't? But if I said "I believe in God," I really think I would be giving the listener a false impression. Look, *everybody* believes in the ultimate reality—which is nothing more than the sum total of all there is. But would you not say that some people are atheists and some are not? Or do you believe that everyone is in fact a theist, whether he calls himself one or not?

Theist: That is a very difficult question to answer.

Naturalist: It seems to me that the phrase "ultimate reality" is absolutely neutral, whereas the word "God" is psychologically highly loaded. I believe that to switch to the word "God" would in fact amount to far more than accepting a nominal definition—it would probably change my entire emotional attitudes toward life.

Theist: And is not the word "Nature" also emotionally loaded?

Naturalist: Of course it is! When I say "everything is Nature," I am saying far more than if I were to utter the silly tautology "everything is everything." The statement "everything is Nature" does not pretend to be a statement of *fact,* a logical or scientific "truth," or to be positivistically meaningful. It is simply the expression of a feeling which is essentially mystical.

Theist: And does not the expression "God is everything" also express a mystical feeling?

Naturalist: Yes it does, but not the same mystical feeling which I wish to express. I think this is precisely the reason we are each so reluctant to change our vocabularies.

I believe the last statement of the naturalist is absolutely right. The naturalist and the theist really have different *purposes,* and each uses the terminology best suited to his purpose. Uniformity of vocabulary is of course important to avoid ambiguities in conversation. But speech and thought have far more purposes than just to avoid ambiguity.

Pragmatism and Truth

☐ I am certainly no pragmatist. Indeed, the idea strikes me as most odd that because a proposition *seems* true to one and "works" for one, that therefore the idea *is* true (or is true for the one). So far I have *stated* that I disbelieve pragmatism, but what *proof* do I have that it is false? Good question; what proof *do* I have? Well, my favorite refutation of pragmatism (not that I don't have others) is, of all things, on pragmatic grounds. Pragmatism *seems* false to me, and has never worked for me, ergo on pragmatic grounds I am perfectly justified in calling it false. That's the end of my argument! With it, I have silenced every pragmatist I have met.[1]

There is another form of pragmatism (which I am not absolutely sure should be called "pragmatism," but which at least has strongly pragmatic overtones) which I now wish to consider. I am thinking of the position which does recognize absolute truth and absolute falsity but which nevertheless believes that it is all right to embrace a false belief provided the false belief proves beneficial to the believer as well as to other people around him. I am thinking of the type who says, "Of course I believe so-and-so's beliefs are *wrong*, but he seems ever so much more happy and helpful since he has adopted them, so perhaps it's a good thing he believes those things after all." Now the opponents of this position are usually dedicated to the shrine of truth; they believe that the pursuit of truth is the most important thing in life and that one should *never* compromise with the truth for the mere sake of its possible utility.

My own position on this question is sort of midway. The main purpose of this essay is to point out to the critics of this type of pragmatism a certain interesting feature which they have possibly not considered.

[1] Actually, I have never met any pragmatists, hence my statement is vacuously true. But I'm sure if I did meet some, it wouldn't be any different.

All right, then, let me fully agree with the critics that the pursuit of truth *is* the most important thing in life. Let me take even a possibly exaggerated position on this and go so far as to measure an individual's worth by his capacity to know the truth or better yet to discover new truths which are of value to others. Does it follow from this that it is therefore undesirable that such an individual believe something false? My answer is, *not necessarily!* Suppose the false belief does not merely have utility, but actually enables him to *know many other valuable truths which he would otherwise not know?* This may sound crazy, but I have in mind the following type of situation:

Suppose the man is a scientist, one who has great scientific ability but who gets into some emotional crisis which makes it impossible for him to continue working. He is totally unable to learn, and he is totally unable to discover or create. What should he do? Most people these days would probably suggest that he try to get psychiatric help. (Of course, some extreme anti-pragmatists who also believe that psychiatry itself is based on false premises would not even approve of that, but let that point go). Well, suppose psychiatry turns out to be of no help to him at all; either his analyst is incompetent, or psychiatry itself is not competent to handle his type of problem—or to put it more charitably to the psychiatrists, he is not the type who is amenable to psychiatric treatment.[2] At any rate, psychiatry has not helped this scientist. Then out of the blue, the scientist has a fantastic religious conversion *but happens to embrace a religion which is totally false!*[3] But this false religious belief totally restores his former serenity, and he

[2]This may be an unwarranted assumption. I know at least one psychiatrist who believes that *everyone* is amenable to treatment.

[3]Again, a possibly unwarranted assumption. I question whether any religious beliefs can be correctly characterized as either true or false. Some logical positivists will agree to this and characterize such religious beliefs as "meaningless." But I question that too! I don't believe that for a sentence to be meaningful it must be either true or false. But this is an extremely subtle matter which would take me far afield.

is accordingly able to work again. As a result, he discovers valuable scientific *truths* for the rest of his life.

Now, how do you worshippers at the shrine of truth react to this situation? We have here the curious case of a person who discovers many important scientific truths by virtue of holding a belief which is false! So is it or is it not better that he hold this false one?

Some might reply, "Well, in an exceptional case like this, all right, but surely there should be some *saner* method of avoiding emotional disturbance than by embracing false beliefs." I quite agree with this. If a saner method can be found, I would of course prefer it. But what if a saner method cannot be found?

Another objection to my point of view was recently raised by someone. This objection is really quite silly, but I think I should briefly consider it on the off chance that some readers may suffer the same misunderstanding. The objection is: "How can anyone in his right mind deliberately try to believe something which he presently disbelieves just because he believes it would be beneficial to do so?" As I say, this represents a total misunderstanding of what I am advocating. Of course I don't believe that a person should, or even could do anything that foolish! I am not advocating that you try to believe things you regard as false just because it will help you. This essay is merely a plea for tolerance toward others who believe things *you* regard as false. More strongly, I am saying that even if your most cherished ideal is the pursuit of truth, you should realize that if someone else believes something you regard as false—*or even if the belief really is false*—this belief may well free him to have other beliefs which are true and which otherwise would not have been accessible to him.

Moral Responsibility

☐ There are some who claim we are "morally responsible for our acts" and who feel it to be of the utmost importance that we recognize and acknowledge our responsibility for the things we do. Then there are those who believe that free will and choice are only illusions and that we are *not* responsible for the things we do.

In my opinion, both points of view are equally horrible, equally damaging, equally crippling—at least for many people. Suppose one is told, "You are morally responsible for your acts." Then he may very well feel extremely guilty, anxiety ridden, worried as to whether he will "choose to do the right thing," and more generally uncertain as to whether he has any right to exist. On the other hand, if he is told, "You are not responsible for your acts," he will at best find these words shallow, hollow, and totally unconvincing, or more likely still, will feel highly insulted and resentful at being spoken to as if he were a mere mechanism, and will suffer a feeling of terrible impotence as if he somehow were incapable of having a say as to how he should manage his own life. So in short, to be told we are morally responsible only induces harmful guilt, and to be told we are not, only induces impotence. And since guilt and impotence are both bad (in my opinion), I am therefore against both the affirmation and denial of moral responsibility.

I also wish to say something about the one who affirms or denies moral responsibility. As I see it, anyone who tells people they are morally responsible is an ego-assertive sadist, and anyone who tells them they are not is an idiot. Also anyone who holds himself morally responsible is a masochist, and anyone who says to himself, "I am not morally responsible" is only trying to relieve his own guilt feelings *which should never have been there in the first place.* Thus I am strongly against both the affirmation and denial of moral responsibility.

There is one natural argument against my position which at once leaps to the mind: Just because I personally react adversely to being told either that I am or that I am not responsible, does not mean that others react the same way. Why then do I project my problems on others? I can only answer that if you feel better off thinking of yourself as morally responsible, then by all means do so. Or if you feel better off thinking you are not responsible, then again by all means do so. I am not against a person's having either of those two attitudes about the conduct of his own life. (Indeed, I suspect that if a person takes either of these two attitudes, he knows what he is doing and probably has a sound psychological reason.) I am not against having these attitudes for oneself, but only against telling *other* people that they are morally responsible or that they are not. What I have just said is not quite right; it is silly for me to reprimand those who tell others what they really believe. So rather than say that people *shouldn't* tell others that they are responsible or that they are not responsible, let me say that I simply feel sorry for, and wish to give moral support to those who, like myself, have been told such horrible things.

I am, of course, here considering the question of moral responsibility from the viewpoint of its helpfulness or harmfulness rather than from the viewpoint of its truth or falsity. Volumes could (and indeed have) been written on the question of whether it is true or false that we are morally responsible. Let me say at once that I regard this question as unanswerable (though I think it is not without interest to try to answer it). Indeed I, like many logical positivists, suspect that the question may well be a pseudo-question. I believe the statement "We are morally responsible for our acts" is in itself neither true nor false, but only a reflection (and an important one) on the state of mind of the one who utters it. Nevertheless, *feelings* of moral responsibility are very important and very real. I know these feelings are real, since I have experienced them. And I believe these feelings are very bad. Likewise, "feelings" of not being responsible are very real, and, for reasons I have already given, equally bad.

Let me tell you an incident: I know one computer scientist who takes a completely deterministic attitude toward the entire universe including human beings. He and I were once discussing Skinner, and he said: "Of course we all know that free will and moral responsibility are mere fictions. But they are useful fictions and should be encouraged." My thought to this was: "How horrible! I would never encourage belief in what I held to be a fiction, regardless of how useful it might be." But what I *replied* was: "I do not believe free will is a fiction. As to moral responsibility, I am not against it on any scientific or philosophical grounds, but, curiously enough, on moral grounds. I simply do not think it is *nice* to hold people morally responsible." He thoughtfully replied, "That's interesting."

Whenever anyone tells me that I am morally responsible, I argue that I am not. And when anyone tells me I am not, I argue that I am. This is not due to mere perversity on my part! I merely try to restore the balance. If someone forced me to drink excessive alkaline, I would immediately try to neutralize it by drinking acid, and vice versa. So when anyone tries to put poisonous ideas into my system, I immediately try to neutralize them by ingesting the opposite kind of poison.

The whole question of moral responsibility is simply a nightmarish duality. How wonderful to be like some people I have known who are free of all this! They act as freely as birds in flight, without being fettered by any feeling of "being responsible" nor by any feeling of "constraint by past causes."

I can think of no better conclusion to this essay than to quote the following passage of Chuangtse, which says even better what I am trying to say:

> Unawareness of one's feet is the
> mark of shoes that fit.
> Unawareness of one's waist is the
> mark of a belt that fits.
> Unawareness of right and wrong is the
> mark of a mind at ease.

Postscript: In this essay I have tried my best to expose the notion of "moral responsibility" for the nightmarish duality that it is. It is really amazing—and frightening—the number of questions which have been raised which logically speaking seem to demand a "yes" or "no" answer but which are really such that both answers are somehow wrong, and more significantly still, are both extremely harmful. One such question is, "Am I morally responsible?" Another such question (which seems to me less damaging, however) is, "Do we have free will?" Interestingly enough, Nietzsche in his *Beyond Good and Evil* says much the same thing about free will as I have said about moral responsibility. Here is what Nietzsche says:

"If anyone should find out in this manner the crass stupidity of the celebrated conception of 'free will' and put it out of his head altogether, I beg of him to carry his 'enlightenment' a step further and also put out of his head the contrary of this monstrous conception of 'free will': I mean 'non-free will.'"

This passage really strikes me as the most sensible thing I have ever read on the question of free will. Nietzsche later makes the following remarks:

"And in general, if I have observed correctly, the 'non-freedom of the will' is regarded as a problem from two entirely opposite standpoints, but always in a profoundly *personal* manner: some will not give up their 'responsibility,' their belief in *themselves*, the personal right to *their* merits, at any price (the vain races belong to this class); others on the contrary, do not wish to be answerable for anything, or blamed for anything, and owing to an inward self-contempt, seek *to get out of the business*, no matter how. The latter, when they write books, are in the habit at present of taking the side of criminals; a sort of socialistic sympathy is their favorite disguise."

Can God Be Stubborn?

☐ In Mackay's stimulating article "Brain and Will"[1] he considers the following problem. Suppose that the universe is

---

[1] D. M. Mackay, "Brain and Will," *Listener*, May 9, 16, 1957 (reprinted in *Faith and Thought*, vol. 90, 1958). Also reprinted in *Body and Mind*, ed. G. N. A. Vessey (London: George Allen and Unwin, Ltd., 1964).

totally deterministic; all phenomena, including personal decisions, are completely determined in advance by, say, purely
physical laws. Would it then, in principle, be possible to
predict correctly another's behavior? The answer is *no*, as his
following example shows. Consider a subject who is asked to
choose between porridge and prunes for breakfast. Suppose he
chooses, say, prunes. Now, present is a super-physiologist who
perfectly knows the present brain state of the subject as well as
the present configuration of the entire universe. The question
is, can the physiologist predict *to the subject* what the subject
will choose? If the subject is *stubborn*, then the answer is
clearly no. The subject might well say, "If you tell me that I will
choose prunes, then I will choose porridge; if you tell me that I
will choose porridge, then I will choose prunes." Under these
circumstances, it is *logically impossible* for the physiologist to
predict correctly to the subject what he will do.

This, of course, does not mean that it is impossible for the
physiologist to *know* what the subject will choose; it is just
that he cannot correctly *tell* him. The physiologist could, for
example, write his prediction on a piece of paper and refuse to
show it to the subject until after he has chosen. At any rate, the
subject's choice will depend on what the physiologist first
does—i.e., on whether he predicts "porridge," "prunes," or
remains silent. So we might further observe that in order for
the physiologist to predict *even to himself* what the *subject* will
choose, the physiologist must first know what *he* will do.

This, then, suggests the following variant of the situation.
What happens if the physiologist and the subject are the same
person? Now the fun begins! Suppose the physiologist knows
that in five minutes his wife is going to ask him whether he
wants prunes or porridge for breakfast. Now, the physiologist
knows so much about the universe, and is so clever and facile,
it should take him only three minutes to compute what he
shall choose. But suppose the physiologist is also very stubborn and decides, "If my calculations come out that I will
choose prunes, then I will deliberately choose porridge, and
vice versa." Under *these* conditions, it is logically *impossible* for

the physiologist to predict even to himself what he will choose.[2] On the other hand, if the physiologist is *not* in a stubborn frame of mind, then it may in principle be possible for him to predict what he will do. So whether he can make such a prediction or not may depend *solely* on his decision whether to be stubborn or not! In other words, it is in *his power* to determine whether he can predict the future or not. But, for all we know, the world *may* be determined, in which case it is determined whether he decides to be stubborn or not. I could carry this analysis still further, but I am getting a little weary!

Let me now return to the case where the super-physiologist decides to be stubborn, and *is* stubborn. Then, as we have seen, he *cannot* predict his future. Assuming still that the universe is deterministic, this means that he is *not* capable of knowing all the laws of the universe—in other words, he is not omniscient. So the upshot of our argument is that it is logically impossible for a being to be both omniscient and stubborn. From which a theologian might be happy to draw the conclusion "God is not stubborn."

In conclusion, let me remark that the whole question of "determinism" is really not central to our main point. This point can be stated without reference to determinism at all; indeed, it can be stated in the simple truism, "A being cannot predict what he will do if he refuses to do what he predicts."

Afterthought: I can imagine an omniscient being who, for some strange reason, *wants* to be stubborn, but who also does not want to lose his omniscience, and who realizes that he cannot have both. I can just imagine such a poor being saying, "Gee, I so *want* to be stubborn, but I certainly don't want to lose my omniscience; what can I do?"

[2]Yet the physiologist might *know* what he will choose. He might, for example, have an *appetite* for prunes, and none for porridge, and so he might damn well *know* he will choose prunes. But "know" in this sense is very different from ability to predict in the sense of *calculate*.

Vignettes

A Thought on Mysticism

☐ I think mysticism might be characterized as the study of those propositions which are equivalent to their own negations. The Western point of view is that the class of all such propositions is empty. The Eastern point of view is that this class is empty if and only if it isn't.

Eastern and Western Philosophy in a Nutshell

☐ *Easterner:*And therefore our supreme aim is to understand and emancipate the mind from the troubles and anxieties of life.

Westerner (anxiously): Oh no! Troubles and anxieties have *survival* value!

Another Thought on Mysticism

☐ I have sometimes wondered if mystics—or rather mystic aspirants—are not in quest of something which most people already have?

A Remark on Kant's Categorical Imperative

☐ It recently occurred to me, quite to my surprise, that in rejecting Kant's Categorical Imperative,[1] I am implicitly obeying it, since I do indeed will it as universal law that everybody reject the Categorical Imperative.

Magic Objects

☐ Wouldn't it be funny if things were really such that physical objects existed only when they were *not* perceived! That is, while they were not seen, felt, heard, etc., they existed perfectly well, but the minute one perceived them, they went out of

---

[1] Which is the principle that one should perform only those acts which one wills as universal law that everyone perform.

existence; they would then *appear* to exist, but the appearance would be only an illusion.

The funniest thing of all is that such a universe is logically possible!

A Remark on Wittgenstein

☐ In the Tractatus occurs the well-known passage, "My propositions are elucidatory in this way: he who understands them finally recognizes them as senseless...whereof one cannot speak thereof one must be silent."

Many have complained that the first sentence is *nonsensical*. Of course if it is taken *literally*, one can reason that if a proposition is understandable, then it cannot be senseless, and hence the sentence itself is contradictory. But why take it so literally? Rather, I would interpret it to mean "The purpose of these propositions is to induce certain psychological changes in the reader, in particular to induce certain critical attitudes. If the reader does adopt these critical attitudes and applies them to these very propositions, then the propositions will have served their purpose."

This is rather reminiscent of the joke about a merchant who wished to sell fish heads to a customer and told him that fish heads are great brain food, that they make the eater smart. And since they have *such* marvelous value to the brain, they are extremely expensive. The customer then bought one for an exhorbitant price and ate it on the spot. After a few moments of reflective silence, the customer said, "It just occurred to me, what sort of a deal was this anyhow? At the market one can buy a whole fish for far less than you charged for just the head!" Upon which the merchant said, "See, it is already making you smart!"

Is Science Incompatible with Teleology?

☐ The biologist Jacques Monod has said: "The scientific

attitude implies what I call the postulate of objectivity—that is to say, the fundamental postulate that there is no plan, that there is no intention in the universe."

This viewpoint strikes me as ridiculous! I am not asserting that there *is* a plan in the universe, but only that there could be, and the existence of such a plan is neither implied by nor in conflict with Science as we know it.

Let us consider the following concrete situation: An earthquake occurs in a small village, killing and maiming many of the inhabitants. In this village there are both priests and atheistic scientists—geologists, let us say. The priests point out to the survivors what extremely sinful lives the inhabitants had been living and assert that the earthquake was a punishment from God—or alternatively, a lesson from God to the inhabitants to mend their ways. The geologists of the town laugh at the "superstitious" priests and assure the survivors that the sinful lives had absolutely nothing to do with the earthquake, offering them a purely rational geological explanation instead. So in short, the priests claim that the sinning of the inhabitants was the *cause* of the earthquake, and the geologists claim that the sinning had absolutely nothing to do with the earthquake. Taking both their claims literally, it seems to me that both could be wrong, but if both claims were changed just a little (in a manner I will later indicate), then although they would still superficially appear to contradict each other, I claim they could both be right.

What I have in mind is this: It seems logically possible that eons ago God planned the laws of physics in such a way that the universe could then run by itself without any further intervention, and without any "miracles." But the laws were so planned that earthquakes would occur most often in places which were desolate or inhabited by "sinful" people. This may sound ridiculous, but it is logically possible. It would involve fantastic ingenuity to plan such a universe, one in which everything which does happen "should" happen, but such a universe is not self-contradictory. Therefore it is possible that this universe is such a universe. (Even without a God, it is logically possible—though fantastically improbable—that this universe is such a universe.) So let us suppose that this universe is such a universe. Then the priests of our village would not be quite right in saying that the sinning of the people *caused* the earthquake; rather it would be that the sinning of the people and the earthquake had a common cause (way back!) and were, so to speak, *synchronized*. It is because of this synchronization that the geologist would not be right either when he said that the sinning of the people had "nothing to do with the earthquake." (True, the sinning did not *cause* the earthquake, but it had something to do with it in the sense of synchronization.) Thus as I say, in such a universe, the priest and the geologist are both wrong. But now, suppose they both modified their statements as follows: The priest says, "The *purpose* of the earthquake was to reform the surviving inhabitants," and the geologist says, "The *cause* of the earthquake was such and such unstable formation." Then I say, both could be right. To repeat my point, God could have seen to it in advance that unstable earth formations, or other such things, would be perfectly correlated with sinful living.

I am not saying that I believe our universe is in fact like this. But I believe that it *could* be and that there is no logical incompatibility between a purely causal and a purely teleological explanation of things happening as they do. The particular example I chose, though illustrating well enough the principle of syncretism (carried to a theological extreme), was a very poor one from other points of view. In the first place, I do not believe in any such thing as "sin"; I chose this merely because it has figured so prominently in so many religiously oriented teleological explanations of the past. Second, I think earthquakes would be a most inefficient "sign from God" that people had best change their ways. But the important thing is not the peculiarities of the example I chose, but the idea of synchronization which it illustrates.

I think that some of the aversions one may have to teleology are due to some of the hideous teleological explanations given in the past, e.g., that the purpose of syphilis is to "punish sexual immorality." Indeed, there is much teleology in the Bible which is simply not acceptable to our present ethical tastes. So by all means, then, let us reject such and such particular teleological explanations, but this does not mean that we must throw away teleology altogether!

The reader might ask, "But why *should* I accept teleology?" My answer is that there is absolutely no reason why you *should*, nor do I believe that you should. The fact simply is that some people are so constituted that they can no more conceive of there being no purpose in the universe than others can conceive of things happening without a cause. Scientific thinking and teleological thinking are both extremely fundamental to human nature. My purpose is not to propagandize for teleology, but merely to point out that the two types of thinking are nowise incompatible.

I cannot help recalling that when Laplace finished his *Celestial Mechanics*, some woman criticized it on the grounds that in it he had made no mention of God. Laplace answered, "But I had no need of such a hypothesis." He was, of course, perfectly right. From Newton's laws, a great deal could be explained about the motions of heavenly bodies, without any appeal to anything like "God" or "purpose." The laws are indeed sufficient unto themselves (insofar as they are accurate). But this does not mean that there is no *purpose* for the planets moving as they do, or for the laws being what they are. Indeed, the laws themselves cannot constitute the slightest confirmation or refutation of the proposition that the laws themselves have a purpose.

Mysticism and Color Vision

☐ In Chapter VII of Bertrand Russell's book *Religion and Science* [1]—the chapter entitled "Mysticism"—one of the things

[1] New York: Henry Holt and Company, 1935.

Russell considers is how one can objectively test the validity of certain insights claimed by mystics. To quote, "As men of scientific temper, we shall naturally first ask whether there is any way by which we can ourselves obtain the same evidence at first hand." Russell then goes on to consider the claim of some that by practicing certain yoga breathing exercises, one would then obtain firsthand direct perception. But then Russell doubts that upon returning to a normal mode of respiration, he would be sure whether the vision was to be believed. It is this doubt which I wish to discuss.

If this doubt is justified, then it seems to me that there can be *no* valid way of investigating the claims of mystics. If the mystics are right (which I am not claiming), then it seems that there is no possible way someone like Russell could ever know it! And it seems to me lamentable that skeptics should be forever cut off from a source of possibly significant truth.

Later on in the chapter, Russell discusses effects of nitrous oxide and compares them with effects of alcohol. As he points out, the drunkard who sees snakes does not imagine afterwards that he has had a revelation hidden from others. Indeed it is true that most alcoholics do not believe in the reality of their delusions but recognize them as sheer delusions (sometimes even when they are having them). But it appears (as related by William James and others) that with nitrous oxide, people come out of it with a strong feeling and conviction that the intoxication has produced highly significant revealed truths which are normally inaccessible. However, Russell claims that from a scientific point of view, no distinction should be made between the effects of alcohol and those of nitrous oxide. Russell rejects them both on the grounds, it appears, of distrust of any so-called "insights" incurred under an *abnormal* state. Now is it really scientifically sound to reject insights incurred under an abnormal state on the mere grounds that the state is abnormal? This is the second question I wish to consider. I can well understand the fear (which I share) of subjecting oneself to an abnormal state for the purpose of obtaining insights. But this is very different from *rejecting* the insights of others who have been in such a state.

At this point I wish to consider an important analogy. Imagine a planet in which all the inhabitants suffer from some chemical deficiency, and as a result are all color-blind. Now consider that someone takes a drug, which among other effects, temporarily removes this deficiency, and so the taker, for the first time in his life, sees the world in all its glorious colors. And as the drug gradually wears off, the world gradually returns to its drab black and white. Now suppose the investigator is of a critical and scientific turn of mind. What should he conclude? Should he reject this experience as hallucinatory? After all, it was incurred in an abnormal state (abnormal, that is, relative to the other inhabitants of the planet). And how would others react to his account of the experience? How could he even describe the experience? What words can convey color sensation to one who has never experienced it? He would have to use such mystic terminology as "ineffable."

Suppose further (to complicate matters) that this drug, in addition to inducing color vision, had horrible side effects, like inducing frenzy, violence, and murder (which sometimes happens, e.g., with LSD). So suppose some people took the drug, and a few committed such acts of violence. Imagine the popular reaction! It would probably be something like, "If this thing called 'color vision' has to be accompanied by acts of violence, we can certainly live without it!" I myself would heartily agree. But I think a more *fruitful* attitude would be to try to discover a method of universally obtaining color vision without these ill side effects.

I wonder if an inhabitant of this planet who had once seen color *could* be capable of doubting it once the vision wore off. Maybe Russell is simply wrong in claiming that if he once had mystic vision, he would disbelieve it after it was over. It may be that such vision has as psychologically compelling an effect as color vision.

145

Of course my analogy (like most analogies) has several weaknesses. For one thing, an inhabitant of a planet who was the only one to possess color vision could certainly demonstrate that he had *some* power not possessed by the rest. For example, he could paint one box red and nineteen others blue (of the same value), have an object placed in the red box, close all the lids, invite the company to shuffle around the boxes while he was out of the room, and upon returning could immediately point to the box containing the object. (By contrast, the mystic cannot demonstrate his insight to anyone not already sharing it.)

I believe the following variant of the analogy is still better. Suppose that on this imaginary planet, color vision was induced not by a drug, but by some extremely rare and improbable (and totally unknown) concatenation of physical and psychological circumstances which induced the state for a short time (say a few hours) and which in all likelihood would never occur again. Or alternatively, suppose that on this planet, color vision came by evolution, and that the first few people who had it had it for only a few hours, and only once in their lifetime. At any rate, whatever the causes might be, let us assume that the *effects* are that for some totally unknown reason, a *very* few men had color vision for a short time and only once. Imagine now the state of one of the early ones—say the first—to have it. Since he could not induce it a second time, he would have even more grounds for suspecting it was an illusion. Yet *could* he really believe this? And how would others react? This time the man could not even convince others that he had some unusual power, since he could not reproduce the phenomenon. Most people would probably think he was simply a crackpot. A few romantically (and perhaps unscientifically) minded souls who were looking for something new and sensational would probably have "faith" that this man had really seen something new and wonderful. Now the man would have to invent some new word, say "color," to describe that new marvelous property he perceived in physical objects. But he could give absolutely no definition at all of what he even *meant* by this word. All he could say is, "I saw *something* marvelous, but I cannot *describe* to you what it is. I saw it with my very own *eyes*, but it is not a size nor a shape, not a texture, but something inconceivably different, something for which I use the word 'color.'" Most likely, the more positivistically minded philosophers of the planet would reject the word "color" as simply meaningless. And the statement "objects have color," which the skeptic would believe as *false* and the romantic would have faith in as *true*, the positivist would reject as neither true nor false, but simply as *meaningless*, since the very word "color" had no well-defined meaning.

I think one should also be open to the possibility that a mystic might simultaneously perceive things which are real and things which are not. Let me illustrate with an example. An acquaintance of mine once told me that after having taken LSD he saw rocks contracting and expanding. Now this is obviously a hallucination; we all know perfectly well that rocks simply do not expand and contract. Now consider the case of one on my imaginary planet who took a mixture of a color-vision drug with some hallucinogen like LSD. He would then report something like, "I saw the rocks contracting and expanding, and at the same time they had some totally new and wonderful quality which I call 'color.'" So this would be a good case of seeing something real (but extremely uncommon) and something hallucinatory at the same time.

I now wish to return to the question as to whether it is scientifically sound to reject mystic experiences on the grounds that they are incurred in states which are abnormal. The use here of the word "abnormal" I find disturbing. This word, I am afraid, only tends to bias the issue, since it obviously has pathological overtones. What really is abnormal? Persistent use of drugs might indeed by classified as abnormal on the simple grounds of being highly deleterious to physical and mental health. What about yoga breathing exercises; do they adversely affect health? This is extremely doubtful; indeed, from most available reports, it seems to be the very opposite. So should we call this state "abnormal"? It certainly might be termed "unusual" (at least in our Western culture). Now to go one step further, what about mystic insight obtained not by drugs, not by breathing exercises, but by so-called exercises in "meditation"? Of course it might be argued that prolonged meditation leads to a sort of "hypnotic" state, and so the question is really whether hypnotic states should be regarded as abnormal. I guess I would call them abnormal, even though they are (in general) not pathological. But then it should be realized that people in hypnosis most certainly *do* have access to much real information which is not normally accessible (e.g., deeply buried childhood memories).

We now shall go one step further. I recall some Zen authority as having claimed that the state of Zen "enlightenment" meditation exercises is virtually identical with that arrived at by Immanuel Kant as a result of years of *philosophizing*! And so I ask, what about philosophizing; is *this* abnormal? Surely, most people think of philosophers as pretty balmy! And although I do not think of the state of mind of philosophers as pathological, surely there is *some* sense in which it can be honestly described as not *normal*! It seems to me that years and years of philosophizing may well have psychological effects fully as profound as any meditation, breathing exercises, or drug taking.

My personal attitudes to all this are as follows. I believe drug taking *does* produce *some* valid insights, but I would not dream of personally trying the experiment. Breathing exercises I would trust *far* more, but perhaps not enough to actually do them. Meditation I trust still more, and this I might consider. Philosophizing I trust still more. Most of all, I value those mystic insights which simply come out of a clear blue, without the slightest attempt at gaining them!

THE FRUIT OF KNOWLEDGE

Interestingly enough, this is precisely what happened to Bertrand Russell! He describes this perfectly in the first volume of his autobiography. His mystic feelings just came upon him, he knew not how. And it is of interest that the quality of his letters during this period is so much more beautiful, sensitive, and profound than that of any of those written earlier or much later.

VIII PLANET WITHOUT LAUGHTER

Planet Without Laughter

☐ I. *The Modern Period*

Once upon a time there was a universe. In this universe there was a planet. On this planet there was virtually no laughter. Nothing like "humor" was really known. People never laughed, nor jested, nor kidded, nor joked, nor anything like that. The inhabitants were extremely serious, conscientious, sincere, hard-working, studious, well-wishing, and moral. But of humor they knew nothing. All except for a small minority who had *some* feeling for what humor was. These people occasionally laughed and joked. Their behavior was extremely alarming to everyone else and was regarded as an obviously pathological phenomenon. These few people were called "laughers," and they were promptly hospitalized. What was so alarming about their behavior was not only the strange noises they made and the peculiar facial expressions they bore while "laughing," but the utterly pathological things they said! They seemed to lose all sense of reality. They said things which were totally irrational, indeed sometimes logically self-contradictory. In short, they behaved exactly like anyone else who was deluded or hallucinated, hence they were put into hospitals.

Medical opinions differed as to the cause of this "humor" disease. Some doctors believed it was organic, others that it was a functional disorder; some thought it was due to a chemical imbalance, others claimed that it was purely psychogenic and often caused by a faulty environment. Indeed, to support the claims of the latter, it was definitely verified that this "laughter" was somewhat contagious and that certain individuals became laughers for the first time in their life only after repeated contact with other laughers. Indeed, this was another thing which made the laughers so very dangerous; they were not only hallucinated themselves, but tended to cause these hallucinations to others! Hence they had to be hospitalized not only for their own sakes, but also for the sake of society.

At any rate, the well-known phenomenon of "contact laughter" added much support to the theory that laughter was of psychogenic origin. But unfortunately, no psychiatrist who held the functional theory and who applied it in the treatment of laughter patients had any therapeutic results. Not a single laugher was ever cured by purely analytic means. On the other hand, those psychiatrists who used chemical therapy had spectacular results! One drug, called "laughazone," was particularly miraculous. It was best administered intravenously, though it could also be used orally. The effects of only one dose usually lasted six or seven months. Almost immediately upon administration, the patient would stop laughing as well as stop this verbal activity called "joking," and instead would start screaming. The screams would mount to a violent and agonizing pitch within about twenty minutes and would continue at this pitch for virtually the whole of the six- or seven-month duration. The patient would just lie there screaming hour after hour, day after day, week after week, and month after month. And the most amazing thing of all is that not once during this screaming period did the patient ever laugh or crack a joke or even smile. Yes, this drug was really phenomenal!

Yet not all the doctors were wholly satisfied. Some took the poistion that the side effects of this drug—namely the screaming—might be even more damaging to the patient than the original laughing. They pointed out that the patient appeared "happier" as a laugher than as a screamer. The opposition granted that the patient was happier in the original state of laughter than as a screamer, but on the other hand, the patient in the screaming state was less deluded or hallucinated than in the laughing state. They said, "What use is it to be merely happy, when the happiness is based purely on psychotic delusions? Is it not better to be rid of these delusions, even if the process is painful? After all, who wants to live in a fool's paradise?" This was a difficult argument to answer! Yet some of the doctors preferred to see their patients in their happier, more natural states of humorous psychotic delusion than in the more reality-oriented screaming states which appeared to be so unbearably painful.

Just how this drug laughazone worked was a problem which was never satisfactorily answered. All that was known for sure is that it *did* work. Of course there were many conflicting theories, but none of them was ever fully substantiated. One theory claimed that the laugher before treatment was living largely in a fantasy world—indeed his whole trouble was that he often confused fantasy with reality. But curiously enough, the pathology of the laugher made this confusion seem pleasant rather than painful. In other words, the laugher actually *enjoyed* this confusion of fantasy with reality. Now, what the drug did was to dispel completely all the patient's fantasies. Then for the first time the patient was "deconfused" —he no longer lived in a fantasy world, but saw reality as reality really is. But the real world seen realistically was so much less pleasant and beautiful than the former world of fantasy, that the contrast was unbearable, hence produced the shock which led to the screaming.

This was one theory. Another theory claimed that the drug really didn't produce a cure at all—indeed, that to label it a "cure" was a sham and a delusion. All the drug did (according to this school) was to cause unbearable physical and nervous suffering to the taker, and all the patient was screaming from was the horrible pain induced by the drug. This group claimed that the only reason the patient stopped laughing and joking was that he was in extreme pain. To substantiate this theory, it was pointed out that laughers who were not institutionalized, laughers outside the hospital who got into automobile accidents or incurred other physical injuries, were often known to stop laughing for a while. Indeed, laughers when sick or in any kind of physical pain would never laugh and seldom joke. Also laughers who had just lost a husband or wife or brother or sister or child or very close friend were known to stop laughing for many months. All this evidence seemed to point out that pain, whether physical or mental, somehow seemed antithetical to laughter, and hence by analogy it seemed reasonable to conclude that the pain induced by the drug only temporarily "killed" but did not really "cure" the laughter. The proponents of this theory also proposed the hypothesis that even if a perfectly normal person—i.e., a nonlaugher—took this drug, he would experience terrible pain and became a screamer, and hence that the screaming of the patients had absolutely nothing to do with being "disillusioned" or "suddenly reality-oriented" or anything like that; the screaming was due only to the perfectly normal chemical reaction to the drug. However, this hypothesis was never verified nor refuted, since the screams of the patients were so alarming that no normal person would ever volunteer to try the drug himself. Thus the true action of laughazone remains a mystery to this day.

After the six- or seven-month treatment of the patient, he was, for some unknown reason, terribly run down and in a deep state of depression for several weeks, sometimes longer. After this he gradually convalesced, and his original symptoms of laughing and joking would slowly return. The doctors realized to their sorrow that the cure, though real, was only temporary, and so they put the patient through it again. They said, "Yes, we had best give this treatment again and again until the patient gets cured permanently." Now, usually after about the third treatment—especially when these chemical treatments were combined with psychoanalytic treatment administered during the intervening convalescent periods—a miracle happened, and the patient seemed permanently changed. In the psychoanalytic portions of the treatment the psychiatrist carefully explained to the patient how he had been living in a fantasy world, and how when he started facing reality he would at first find it very painful. And amazingly enough, after about the third treatment, the patient actually agreed that the psychiatrist was right! He said: "I see now that you were absolutely right. I was indeed living in a state in which I constantly confused fantasy with reality, and I moreover believed in the existence of an entity called 'Humor.' Yes, I actually believed it to be something real rather than a mere figment of my imagination. But now I see the light. I realize how in error I have been! These drug treatments have done wonders in making me realize how crazy I have been! Indeed, under this drug I have seen things realistically for the first time; I see now that things are *not funny*! As you anticipated, doctor, my facing reality for the first two or three times was most disturbing. But do you want to know the beautiful thing of it, doctor? I am no longer afraid of reality! After facing it a couple of times, I find it is not so frightening after all! I am now *adjusted* to reality. To tell you the truth, doctor, I don't even think I need ever take the drug again. That's right, I no longer need it! In fact, I'm perfectly confident that I could walk out of this hospital this very day and not even be *tempted* ever again to engage in this pathological activity known as 'humor.' Yes, doctor, I really feel like a new man! Moreover, if I were out of this hospital, I as an ex-laugher could spot other laughers and even potential laughers far better than one who has never gone through my experiences, and I could indeed bring them into the hospital for treatment."

Well, when the doctors heard *this* kind of talk, many of them were delighted and promptly arranged to have the patients discharged. But certain follow-up studies gave the doctors cause for grave concern. For one thing, the ex-laughers never *did* bring in laughers or potential laughers for treatment. Second, there were some pretty reliable rumors that these ex-laughers, although they indeed never laughed or joked in public, did so in private and in a highly clandestine fashion. Also, when they met each other, they would go into huddles which somehow savored of the conspiratorial. And so, many of the doctors framed the hypothesis that perhaps the ex-laughers were not really cured after all, but—of all horrors—only *pretended* to be! In other words, it was seriously suggested that the patients, after about the third treatment, were only *simulating* mental health, and were being, of all things, *insincere*! The reason this hypothesis was so shocking is that insincerity

157

was virtually unknown on this planet. From what little was known about the subject, insincerity itself was regarded as another form of psychosis but one which was exceedingly rare.

The question then arose: What *made* these ex-laughers insincere? A few of the bolder physicians suggested that it was simply that the patients pretended to be well in order to avoid any further painful drug treatments. But that hypothesis was generally rejected. The consensus of medical opinion was that insincerity was never this rational nor premeditated, but was something totally irrational and most likely caused by some chemical imbalance. Indeed, it became suspected that laughazone itself, though temporarily curing the laughing psychosis, might be the very agent which was causing the insincerity psychosis. And so the doctors sadly admitted: "The situation is most depressing! Not only does laughazone fail to provide any permanent cure for laughing, but it seems to have this terrible side effect of producing insincerity!" Some of the ex-laughers were recalled to the hospital and their laughazone treatments were resumed; meanwhile another drug, "insincerezone," was simultaneously administered with the hopes of counterbalancing the "insincerity" effects of the laughazone. But a proper balance never seemed to be struck. During the convalescent periods between drug treatments, the patients were either sincere and laughed, or they ceased to laugh but displayed obvious symptoms of insincerity. In other words, *no* chemical means could be found which would make the patient *sincerely stop laughing!* Various types of cerebral surgery were also tried, but again to no avail! Nothing science could do could make these strange uncanny patients give up humor in a really sincere manner. And so the doctors threw up their hands.

I shall return later to the fate of the laugh-patients.

II. *The Middle Period*

The history of this planet can be roughly divided into three periods: the Ancient Period, the Middle Period, and the Modern Period. The Modern Period contained no literature at all on laughter, except in textbooks and periodicals on abnormal psychology. The Middle Period was chock-full of laugh-literature—indeed this constituted the main writings. This literature contained absolutely no material which contemporary laughers called "funny"; indeed the writings were in a wholly sane, serious, scholarly, and philosophic mood. The writings consisted mainly of analysis and commentary on the ancient texts. Now the ancient writings were totally nonphilosophical; they never spoke *about* laughter or anything like that. The ancient writings were simply what the Middle Period called "funny." These archaic manuscripts contained all sorts of incomprehensible contradictory material called "jokes" or "funny stories." It was the main purpose of the Middle Period to evaluate the work of the Ancient Period. The philosophers of the Middle Period—particularly of the Early Middle Period—actually extolled the Ancient Period and referred to it as a "golden age"—more specifically as "the golden age of humor when men could freely laugh and joke and really enjoy life." These writers kept talking about the decline of laughter as a tragedy rather than as a blessing. They claimed that the ancient writings, despite their apparent irrationality and paradox, really contained an extremely important kind of wisdom. Perhaps "wisdom" (they said) was not quite the right word; certainly this "wisdom" was not the kind of knowledge which could be taught by science or reason. To perceive the value of the ancient writings required a certain almost mystical faculty called "Humor." Furthermore this "Humor" arose curiously enough out of the very paradoxical and allegedly "irrational" character of the ancient writings. In other words (and this is what was found so puzzling!), Humor could not flourish in a wholly serious and rational atmosphere.

The main philosophical problem of the Middle Period was to establish whether this mysterious thing called "Humor" really had objective existence or whether it existed only in the imagination. Those who believed it really existed were called *Pro-Humorists*; those who believed it did not were called *skeptics* or *Anti-Humorists*. Among the Pro-Humorists there raged bitter controversy as to whether the existence of Humor could be established by pure reason, or whether it could be known only by an act of faith. The Pro-Humorists were roughly of three sorts; the *Rational* Pro-Humorists, who claimed that the existence of Humor could be established by pure reason; the *Faith*-Humorists, who believed that reason could be somewhat helpful but that an act of faith was crucial; and finally there were the "Mystic-Humorists" (known in modern times as "laughers"), who claimed that neither reason nor faith were of the slightest help in apprehending Humor; the only reliable way it could be known was by direct perception. Reason, they said, leads nowhere. To believe in the existence of Humor on the mere basis of *authority* means that you obviously don't see it for yourself. To have *faith* in the existence of Humor; on what basis is this faith? Is this faith based on acceptance of authority? Is it based on some sort of *hope* that there really is such a thing as Humor? Is it perhaps that the Faith-Humorists believed that Humor, if it really existed, would be something very *good*, and hence, because of their desire for the good, they took an oath to themselves to conduct their lives as if Humor really did exist? Yes, this seemed to be it. But, as the Mystic-Humorists pointed out, this attitude, though well intentioned, was a sad testimony to the fact that the Faith-Humorists could not see humor *directly*. The Mystic-Humorists kept repeating, "If only you could *see* humor directly, you would not need rational arguments nor any faith nor anything like that. You would then *know* that Humor is real."

This phrase "see Humor directly" was particularly apt to be criticized. The Mystic-Humorists actually said: "Yes, we can see humor in many situations. Life is permeated with humor, if you can only see it." The skeptical Anti-Humorists said, "So, you claim you can *see* humor! Tell me, what color is it?" The Mystic-Humorists laughed and said, "Humor doesn't have any *color*!" The skeptics continued: "Oh, so you can see it only in black and white! Well, then, what *shape* is it?" "It doesn't have any form or shape." "Then I am confused! Is humor visible or invisible?" "Of course it is invisible!" "But I thought you just said that you can *see* it. Didn't you say that you could *see* the humor of certain situations?" "Well, yes, I *said* that, but I didn't mean 'see' in the literal sense of 'see with your eyes.' Ocular vision really has nothing to do with it. I used 'see' in the sense of directly perceive, not see with the eyes. The perception, though as direct as vision, is really through a different sense altogether." "A *different* sense? Which sense is it—hearing? If so, what does humor sound like? Or is it smell or taste or touch or what? With which of the five senses do you perceive humor, or is it a combination of more than one of them?" "No, it is not any one of these five senses, nor is it a combination of them. It is a different sense altogether—in a way, it is a nonphysical sense—we call this sense the 'sense of humor.'" "Good God, you literally mean a nonphysical sense? In other words, you mean it is something occult like telepathy or clairvoyance? But scientific integrity requires us not to believe in anything occult; hence we cannot but believe that this Humor is something totally unreal, a mere figment of the imagination."

In vain the Humor-Mystics protested that there was nothing the least bit occult about humor—indeed the idea that humor was something occult struck them as downright "funny." They laughed at the idea, and said: "If it will help you at all, the very statement that humor is something occult is typical of the type of statement which we label 'humorous' and which makes us laugh. There is absolutely nothing occult about Humor. If once you could only see what humor was, you would realize that it is the most natural thing in the world, and also that it is delightfully pleasant." Another thing, the so-called "Mystical-Humorists" kept claiming that the label "Mystic-Humorist" was most misleading. They claimed that there was nothing at all mystical about humor—even though it might *seem* mystical to those who lacked the immediate sense of humor. They said, "Why not rather call us laughers, which is, in fact, what we are." And so the term "Mystic-Humorist" got gradually replaced by "laugher."

The attitudes of the Faith-Humorists toward the laughers is of considerable interest. Not all of them believed in the authenticity of the laughers. Indeed, many of them were extremely envious of those who could sense Humor directly, hence they simply refused to believe that the laughers could really do this. And in some cases they were right, for rather subtle reasons which will later appear. Now, most of the Faith-Humorists did not take a hostile, skeptical attitude toward the laughers, but believed in them wholeheartedly. They knew that the laughers were in direct contact with that which the Faith-Humorists could only reason about and accept on faith. And so they went to the laughers to be taught. Some of the laughers regarded it as their main mission in life to try to bring laughter back to mankind. They became what was known as "Laugh-Masters" and set up institutions—usually in the mountains or by the seashore—known as "Laugh-Monasteries." To these we now turn.

III. *A Sermon*

The Faith-Humorists come to these monasteries to sit at the feet of the Laugh-Masters in order to learn the holy art of Laughter. The methods of instruction used by the Laugh-Masters varied considerably. There was one famous Laugh-Master, Bankoff, who rejected all orthodox methods and who indeed claimed to have no method whatever.[1] He said, "I think the best method is simply to *amuse* them!" Many of the Laugh-Masters felt hurt that their methods were so flagrantly neglected by Bankoff, but they all had to admit that as a *practitioner*, Bankoff was better than any. Bankoff was what we on this planet would call a "clown." It was not so much his *words* that enlightened people, but his actions. He would act in very strange manners. Sometimes during a serious "laugh-

---

[1] Like the famous seventeenth century Zen-Master *Bankei* of this planet.

sermon" he would suddenly, for no apparent reason, do a series of somersaults, and of the thousands of listeners, one or two of them would burst out into laughter for the first time in their lives. They would say: "Oh, so *now* we see what Humor really is! By God, these somersaults have taught us more than all the books we have read on the philosophy of laughter!" Some of the other Laugh-Masters tried Bankoff's somersault techniques, but for some odd reason they could never carry it off. It somehow fell "flat," and so the other Masters had to return to purely verbal methods of instruction.

Sermons were one of the standard techniques. Sometimes they did some good, sometimes not. I now reproduce one of the most famous sermons:

Oh aspirants to Laughter! You have bravely come long distances to worship at the shrine of humor. But alas, how misguided are your efforts! In the first place you insist on sitting at our feet, and on approaching us in an attitude of reverence. You think of us in some sense as "holy men." And none of our efforts can convince you that the very reverential nature of your approach is the very thing which is preventing you from laughing. If only you could see the *humor* of the situation! You think of laughter as something solemn, and you cannot believe us when we assure you that there is absolutely nothing solemn about humor; humor is almost the antithesis of solemnity. The situation is so strange! On the one hand, you perfectly well know that we are in immediate contact with humor—that we experience it firsthand—and yet your preconceived notions of the "theory of humor" are so strong, so thoroughly engraved in your innermost beings, that you cannot believe the things *we* tell you about laughter. You think that, because you have read all the great books on the philosophy of laughter, you know more *about* laughter than we do—even though you know we can laugh and you cannot. You seem to think that knowledge about laughter is somehow more important than the ability to laugh. And you say that your judgments about laughter are more valid than our own. You keep saying that the ability to laugh does not enable the laugher to know what laughter really *is*—only "analysis," according to you, will do that. You are like certain philosophers of art who feel somehow superior to the working artist and who believe that they really know what art is "all about" more than the artist himself. You are also like some philosophers of science who say, "The working scientist rarely knows what science is actually all about." Or some logicians who say, "Most mathematicians, even though they prove great theorems, don't really know what they are doing." Yes, your attitude toward us is of this nature. You worship our ability to laugh, despite the fact that we tell you that worship is entirely the wrong attitude. We grant that worship might be helpful for achieving other values in life, but laughter can never be acquired through worship. If you could only *laugh* at us instead of worshipping us, you would be on the right track. But you do not even understand what we mean when we say this. You insist that laughter *is* something solemn despite everything we say. You say: "Just because you claim that it is not solemn doesn't mean that it necessarily isn't. It may be that it really is solemn, only you fail to recognize its solemnity." What can we say to you when you talk to us like this? All we can say is: "We grant that we cannot with science and logic prove that humor is not solemn. We just *know* it is not. We are sorry to sound so unreasonable and dogmatic, but all we can do is to assure you that once you have acquired a sense of humor, then you will also know that humor is something which is not solemn.

Let me now tell you in more concrete terms what are some of the errors you make—what are some of the false paths you feel compelled to follow which you so earnestly believe will lead you to acquiring a sense of humor. First of all, almost none of you is able to shake off the completely erroneous belief that it takes grim and determined *effort* to acquire a sense of humor. You all seem to believe that the sense of humor is something that you must somehow earn by your own efforts. You regard laughter as a reward for things you *do*. You also regard laughter as an *act of your own*—as something you actively do rather than as something which happens to you. You may find it hard to believe, but much of our laughter is involuntary. Sometimes we cannot *help* laughing. In some humorous situations, we are, as it were, "overcome" by laughter; the laughter almost comes by itself very much like hiccoughs. At any rate, the sense of humor is not something which you can acquire by your own efforts. The main place where effort comes in is in overcoming your prejudiced beliefs that effort is directly necessary to acquire a sense of humor. We understand the quandary you are in. You say: "Well, if we sit back and do nothing, we do not find ourselves laughing. How then are we to learn laughter unless we make *some effort* to do so?" We admit that this is the hardest question in the world to answer. It seems that you are trapped whichever direction you turn; if you do nothing you do not laugh, and if you do something you also do not laugh. How then are you to laugh? Yes, we perfectly understand your problem, and we wish we could give you a wholly rational answer. But unfortunately we cannot. Perhaps our inability to do so is not too dissimilar to your inability to laugh. At any rate, we cannot help you by answering this question; we can only resort to other methods. One thing, though, we feel will help, and that is to point out how most of the efforts you do make are in the wrong direction. Let me indicate four common false roads.

1. Some of you take an approach which is far too objective and scientific. You read all the literature you can find on the philosophy of humor. You perform elaborate linguistic analyses of what the word "humor" could possibly "mean." You keep looking for better and better defini-

tions of the word "humor." In other words, you are trying to define the word "humor" in terms of other words whose meanings you already know. But this is utterly impossible! The word "humor" is simply *not* definable in terms you already understand. The only way you will ever find out what the word "humor" really means is by acquiring a sense of humor. And for this, science and logic cannot help you in the least. Please do not misunderstand; all this analytic study of humor is of great value for psychology and the theory of knowledge. But it should be undertaken *after* rather than before you have acquired the sense of humor. To undertake the study first will certainly not help you, and may very likely harm you. Why will it harm you? Because the very grim, serious scholarliness of your approach will put you in the frame of mind in which humor is not likely to come to you. Some of you are perhaps amazed at my phrase "come to you"? Well, that's exactly as it is! Yes, to a large extent, humor is something that actually comes to you! This is our whole point which is so hard for you to grasp. If you would only relax, only *let yourself go*, only let humor come to you, then it would. But no, you grimly chase it by your serious studies, and all you succeed in doing is to chase it away,

2. The next wrong notion from which so many of you suffer is that the sense of humor is achieved via *morality*. You have been taught that if you lead a *good* life, then you will be rewarded by acquiring this sense of humor. And so you go forth doing good deeds hoping you will get this reward. But this is totally off the track! We are not against morality—most of us value the ethical life—but we absolutely insist that it has nothing to do with the quest for humor. Where did you get this false notion that you must "earn" the sense of humor by being good? And why do you persist in this belief? Actually the moral quality of the laughers on the whole is not significantly different from that of the nonlaughers. Of course some of us laughers are

very fine people, but others are complete rascals. Morality simply has nothing to do with your problem.

3. Closely related to this is the absolutely ghastly idea some of you have been told that humor can come to you only through all sorts of gruesome ascetic practices. And so you starve yourselves, become sexually abstinent, flagellate and otherwise mutilate your bodies hoping that the intense pain you suffer will bear the fruit of humor. But it never does, and no wonder! The more you pain yourselves, the more impossible it is to enjoy humor. There is one minor exception to this; there is a thing we call "bitter humor," and this does arise in response to painful situations. But this type of humor is comparatively rare and moreover is almost impossible to learn before learning the more normal joyful humor. Yes, humor is sometimes really joyful, and it cannot possibly flourish in the morbid atmosphere of asceticism.

4. The most insidious error of all is to try to learn humor by merely *imitating the outward forms of the laughers*! This error is so subtly deceptive and dangerous, and so difficult for us to correct! You listen most attentively to the sound of our laughter, and then you try to make the same sounds yourself. Some of you are quite good at this acoustical imitation, but you cannot fool us! Even if your imitation were perfect, you still would not be really laughing any more than a parrot is able to understand the language he mimics. You ask us how we know your laughter is not genuine, and we answer, "It simply does not *sound* right." You ask us to be more specific and to "correct" your laughter, or the more scientifically minded among you ask us to give you a precise acoustical analysis of the difference between genuine and imitation laughter. You ask us: "Is the pitch wrong? Is it a question of wrong timing? What *is* it about our laughter which is wrong?" You seem disappointed that we make no effort to answer this sort of question. In the first place, we cannot give you a purely

scientific description of how your laughter sounds unconvincing; an acoustical sine-wave analysis is the last thing
we can give. Perhaps if we put our minds on it, we could
train you to laugh more convincingly, but this would be the
worst possible thing in the world for you! Indeed, if you
could learn to perfectly mimic our laughter, then we would
have practically no means left of knowing whether or not
you had a sense of humor. What you utterly fail to realize is
that it is not the ability to laugh correctly which gives you a
sense of humor, but the very reverse. Once you have the
sense of humor, *then* you will automatically and spontaneously laugh correctly without your having to analyze
how you laugh. Yes, we know that you have fallen under the
spell of many books with such titles as "How to Laugh
Correctly," but we can solemnly assure you that no true
laugher would ever write such a book. Indeed, such books
are totally antithetic to the true spirit of humor. You must
remember that the activity of laughter is only the *outward
form* of Humor; Humor itself is something entirely within
the inner spirit. And you can never attain this spirit by any
amount of imitation of outward forms of behavior. Another
way you try to learn by mere imitation is by this ridiculous
practice of memorizing jokes. In a perfectly laborious and
mechanical fashion you commit to memory thousands
upon thousands of jokes and you think you are thereby
acquiring a sense of humor! You call this activity "studying" —you say you are "studying to acquire a sense of
humor." But these jokes are absolutely pointless for you to
learn until *after* you have acquired a sense of humor.
Without this inner sense, you cannot possibly see the real
point of these jokes. True, even without this sense, you can
understand the situations these jokes describe, but these
situations themselves are totally uninteresting unless you
can perceive the *humor* in them. What is this thing we call
the "humor" in them? Since it is not a color, not a sound,
not a smell, not a taste, not a feel, you wonder what on

earth it could be. Some of you keep insisting that since it is none of these things, then it must be something "mystical," and you cannot believe us when we tell you it is not. Once you can see humor for yourself, you will realize it is something as plain as daylight.

Coming back to the point about joke memorization, we can easily see by the way you tell these jokes that you completely fail to see the humor of them. To put the matter quite plainly, you tell them far too *seriously*. A joke is not something like a solemn liturgical chant; it is virtually the opposite in spirit. You tell a joke—or rather recite it—as if you had just come from a funeral! Again, it is pointless for us to give you an acoustical analysis of what is wrong with the way you recite jokes, we can only say that you should *first* acquire the sense of humor and then the proper way of telling jokes usually comes by itself.

The most serious offenders of you do the following: You combine the two techniques of joke memorization and forced laughter, and you are then sure you have "arrived." But God Almighty, how wrong you are! You first parrot forth your "joke" and then parrot forth your "laughter," and are then *sure* you have a sense of humor! You do not realize that your intense preoccupation with the mere outward forms is the very thing which has prevented the spirit of humor from entering your souls. And furthermore, you will not even believe us when we assure you that you are further away than ever. You get angry and ask us to give you *scientific proof* that you do not yet have a sense of humor. You absolutely refuse to trust your intuition in this matter, and you wrathfully leave our monasteries and go forth into the world claiming yourselves to be "authentic laughers." Nothing sabotages our cause more than this! The skeptics who meet you are almost rightfully reinforced in their belief that humor is something which is "a mere sham and delusion." The Faith-Humorists who believe that humor is indeed real, but who are envious of those who have it, and who believe that no authentic laughers exist anymore, are again reinforced in their beliefs when they meet the pseudo-laughers. Yes, the pseudo-laughers are the major cause of the disappearance of humor from this planet. We try our all to stem the tide; whether or not we will succeed, God only knows.

IV. *The Great Legend*

In addition to the monasteries we have described, there were also the temples—called "Humor-Temples" or "Temples of Laughter." These were located mainly in urban areas. They were very different in spirit from the monasteries. Here the worshippers would gather once a week—on Laughday—to worship at the altar of Laughter. In the Early Middle Period, there was indeed laughter in the temples. The congregation would come and the High Priest would read the ancient texts, and everyone would have a great laugh over them. But as the ages went by, laughter disappeared more and more from the temples (as it did elsewhere in the world), and people turned more and more to the *worship* of laughter. They no longer laughed, but started "praying to Laughter."

During the Middle Middle Period, the great question of mankind was "Why is Laughter disappearing from our planet?" Many hypotheses were offered, but more in the spirit of "legends" than of scientific theories. Many were these legends, and the temples started splitting into groups depending on which legend they held to be the truth. The temples became extremely dogmatic and intolerant of each other, and religious warfare developed, each group of temples fighting for its truth. One legend, known as the "Great Legend," achieved the widest popularity and soon dominated over all the other legends. Here is the Great Legend.

In the days before the Ancient Period, in the beginning, there were only two people in the world. These people— call them Adam and Eve—were brought by the Lord into the world at the same time. They were born on Laughday. They lived in total bliss in the Garden of Laughter. They lived mainly by the streams and laughed with the butterflies and sunbeams. Every day the Lord would visit them in the garden and joyfully fill their souls with His delightful humor. He loved them, joked with them, and laughed with them. Their laughter was divine. And so they spent their days in this paradise for many years, until one day a strange green animal, something like a rat and something like a skunk, with mean, small, close-set eyes, came into the garden. This animal perceived the bliss of the couple and waxed mighty jealous. He said, "I will soon do something about *that*!" and sure enough he did! He approached the couple and said: "How can you two grown-up people live like this all your lives? You are not *children*! How can you idle away all your time by this perfectly infantile *laughter*? Don't you realize there are important *duties* to be done? Yes, it may be *pleasant* for you to fritter away all your time in laughter, but at this rate how will you ever *amount to something*? The Lord has given you the precious gift of life, and all you can do is to betray it in this manner? Shame on you! And the Lord—why does He encourage you in these infantalisms? Why does He persist in his daily visits, telling you all these silly jokes, and keeping you like children? Why is He afraid of your growing up? You have the bodies of adults, but minds of infants. Why does the Lord allow this? What is He afraid of? What is He hiding from you? Why does He pretend to be your friend when He is the very one who is deceiving you and who is preventing you from being true to yourselves and fulfilling your real destinies in the universe? Why do you tolerate this? There is one chink in the Lord's armor by which you can save yourselves. The Lord has given you *free will*, by which you can oppose Him. You can put a stop to this situation; *it is up to you*! Only by your *own efforts* can you prevent the Lord from keeping you in bondage forever."

Thus spoke the Evil Animal. He came back to the garden day after day and slowly but surely beguiled the two Children of Laughter. Now the remarkable thing is that for the most part, neither Adam nor Eve really believed nor trusted the Animal, at least on a conscious level. They somehow did not like the Animal's *looks*; there was a certain expression in its eyes which somehow aroused their suspicions. And yet, the Animal said some remarkable things. In particular, they were absolutely flabbergasted to learn that they had such a thing as free will! Such a strange idea had never occurred to them before. Their lives had flowed by so beautifully, happily, spontaneously, and effortlessly that it never seemed to them as if they themselves were ever *doing* anything. It seemed that things were happening *to* them rather than that they themselves were active agents. For example, when they came within sight or scent of a delicious fruit, it seemed as if the fruit drew them like a magnet, rather than that they *chose* to eat the fruit! To put the matter better, it was not so much that they felt passive rather than active, but rather it was that they never made any distinction between passivity and activity. And so the idea that they could *choose* was a stunning novelty. It gave them an exhilarating sense of *power.* They of their own free will could now *do* things! In particular, they could, if they chose, *amount to something.* The question then arose: *Should* they amount to something? This notion of "should" was also quite new. Formerly, since they had felt that they were merely part of the stream of life rather than actively living it, ethical notions of "should," "ought," "duty," etc., had absolutely no meaning for them. But now they knew better. The troubling question arose: Was it *right* or *wrong* for them to sit by idly enjoying life rather than going out and amounting to something?

Adam and Eve also for the first time began philosophizing. They believed the Animal was right in telling them that they had free will. But the question which most puzzled them was whether they had really had free will *before* the Animal informed them of the fact. If they formerly had free will, they certainly had not *known* that they had. And is it possible to be free without knowing that one is free? In other words, was it really true, as the Animal had said, that God had already given them free will, or was it the Animal himself who caused them to have free will? It seemed likely to them that having free will is really no different from believing that one has free will. They wondered whether there might not be other worlds with sentient beings like themselves, and whether these beings had free will. Adam decided that the most likely answer was that some of the beings had free will and others did not; those that believed they did, did, and those that didn't believe, didn't. They once asked the green Animal, "Do *you* have free will?" The Animal answered: "Of course I do! And you too can have free will if you choose to." This answer puzzled them terribly! They replied: "What? You say we can *choose* to have free will? You mean that having free will is a matter of *choice*?" The Animal replied, "Of course it is." Eve then protested, "But I thought you told us that God has already given us free will." The Animal replied: "In a sense He has, but only in a passive rather than an active sense. God has, so to speak, given you the *potentiality* of having free will. Whether you actualize it or not is up to you. God has given you the *ability* to make choices; He does not *force* you to make them. You can use your free will only if you *choose* to." Adam answered, "But if we can *choose* to, that means we already *do* have free will." The Animal replied, "Yes, it is in that sense that God has given you free will."

All this talk puzzled the couple terribly! And thus the sciences of metaphysics and epistemology were born. Meanwhile, they had on their hands the present moral problem of what to *do* with their free will. Should they or should they not go forth and amount to something? They knew they were perfectly happy in the Garden of Laughter, but was their real role in the cosmos to be happy or to fulfill their duties? They debated this for many weeks, and finally decided to remain in the garden and *not* to amount to something. They decided to trust the Lord and not the Evil Animal. Yes, they finally realized, the Lord is their friend and the Animal their foe. And so one day the Animal came into the garden and Adam said: "You have taught us many wonderful things. You have taught us that we have free will. Whether you have taught us this, or whether by some mysterious power you have *caused* us to have free will, or whether it was God who 'allowed' us to have free will, or whether He 'made us' have free will, or whether it is we who have 'chosen' to have free will, we do not know. We do not understand the phenomenon of free will, but we now know that whatever it really is, we certainly have it. Perhaps we have *chosen* to have it; we really don't know. All we now know for sure is that we in fact do have it. And you are absolutely right that we can now use our free will to reject the Lord and His ways. Yes, we are indeed free to do this. But do you not realize that by the very same token we are now free to reject you? Yes, we now have the power to reject you or the Lord. And it is *you* we have decided to reject! Of our own free wills we thoroughly cast you out of our minds and hearts. We reject you and your ways. We will no longer heed you or your words. We cast you out of this very garden. This garden is *our* property; the Lord has given it to *us*, not to *you*! It is our own private property, and you can no longer be here without our permission. We have so far suffered you here only as a guest. But you are no longer a welcome guest. Begone from the Garden, and don't you ever dare return. If we ever find you here again, we will kill you." The Animal departed without a word, and never returned.

Adam and Eve heaved a sigh of relief. They knew that they had chosen to do the right thing. But alas, their troubles were only beginning! Although on a conscious level they had chosen to accept the Lord and reject the Animal, the poison of the Animal had entered their unconscious souls and became the focus of a vast psychic infection. This infection grew and spread from day to day. Pain entered their souls, and they could not understand why. True, they had decided to remain in the Lord's garden and not worry about amounting to something, but deep down they felt that they *should* amount to something. They became strangely restless and dissatisfied. Troubled thoughts crossed their minds; day by day they became more irritable and depressed. The joy started going out of their lives. They were no longer sure they were even happy in the Garden of Laughter.

Then came the Fatal Day. They had spent a long and restless night of troubled sleep. They both dreamed of going through eons and eons of time and never amounting to anything. In the morning they awoke in a state of complete exhaustion. They sorrowfully went together down to the stream and sat for hours in silent gloomy thought. The Lord came by at the usual hour, and perceived they were sad. He then tried to cheer them up with one of his jokes. At this point, Adam's nerves were strained to the brink, and he angrily snapped at the Lord: "We are not amused! Your jokes are *not* funny!" The Lord gazed at them long and sorrowfully and said: "Very well, then, you have chosen to reject Me, my ways, and my humor. I shall not try to force these on you; indeed I could not even if I would. I cannot *make* you laugh, nor would I if I could. You indeed do have free will, as your great bosom pal has told you. You are perfectly free to reject my humor, and I shall never trouble you with it again. You may as well go forth and 'amount to something,' which is what you deep down really want. Yes, you can amount to a great deal—indeed you can beget an entire race. You will go forth and do this. Only you and your progeny will not have me present to guide you with my humor. I will watch over you, but I will not be with you. Slowly but surely, humor will disappear from the world. Without me present to inject fresh humor into your souls, it will slowly wither away and die. Only after centuries have elapsed, when the world is at its blackest pitch of despair, when you yourselves realize your infinite need for me and my ways, *then* you will find the right way to call me back. And when you *truly* call me back, I will return. Until then, adieu!"

V. *Back to the Modern Period*

Thus the Lord spake, and thus it came to pass. Though laughter disappeared more and more from the world, the people of the Middle Period realized that this was a tragic loss rather than a gain, and they did everything possible to stem the tide. Only at the very end of the Middle Period did it first occur to mankind that laughter, far from being something good, was something totally undesirable. People started saying: "Maybe we should *stop* trying to stem the tide. Maybe the tide is our greatest blessing, although we don't know it. Maybe it is high time that this silly archaic thing called 'Humor' should disappear. Maybe laughter was all right for *savages*, but we are now becoming *civilized*!" Yes, this is how they began to speak. At first they referred to laughter as something "silly," but soon they started using the stronger word "crazy." Then the idea fully occurred to mankind that humor was but another form of psychosis; laughter was a type of psychopathology. Thus was ushered in the Modern Period.

And so we are back to the Modern Period. Typical of this period is the fact that most people do not refer to the Ancient, Middle, and Modern periods by these names, but rather as the "Psychotic Period," the "Convalescent Period," and the present "Sane Period." Yes, the world is now sane; there are very few laughers left. If only it were known how to cure them!

We go back to where we left off in Chapter I. We recall leaving the doctors in a quandary as to how to find the proper balance between laughazone and insincerezone which would make the laughers *sincerely* stop laughing. And the painful fact had to be faced that the laughers were not permanently curable, at least in the foreseeable future. So what was to be done? Here medical opinion split into two divergent camps, and the hospitals split into two widely divergent types. Hospitals of Type I were called "laugh-scream hospitals"; those of Type II, "pure-laugh hospitals." In the laugh-scream hospitals the doctors realized that no patients were permanently curable, hence a patient once admitted was admitted for life. All that could be done was to administer the laughazone treatment over and over again for the rest of the patient's life. The discipline at these hospitals was ironclad; no patient was ever released, and there was to be no letup of treatments. It was to be firmly and painfully realized that although no permanent cure was possible, laughazone did provide a temporary cure, and painful as the cure was, it was better for the patient to face reality and scream than to withdraw into his fantasy world of humor and laugh.

For some unknown reason, the patients at the laugh-scream hospitals did not live very long. Few of them survived the sixth or seventh treatment.

The philosophy of the pure-laugh hospitals was entirely different. They agreed with the laugh-scream hospitals that no laugher was permanently curable, and many doubted that he was even temporarily curable. At any rate, even if he were temporarily curable, was it really worth it? Why not let the patient enjoy his life; was it really all that bad that he had these fantasies? And so, like the laugh-scream hospitals, patients were incarcerated for life. But they were given no laughazone treatments—*nor any treatments whatever*! The psychiatrists at these hospitals said to the patients: "You are incurable; your psychosis is hopeless. There is *nothing* you can do to get better. Therefore, do not *try* to get well; do not *fight* your psychosis, but rather go along with it. In other words, try to become adjusted *within the framework* of your laughing-psychosis. You must learn to live with it. You must learn to *enjoy* your laughter." One patient responded: "But doctor, we *do* enjoy our laughter! We already *are* adjusted to our humor." The doctor, who perhaps had not quite understood him, replied, "No, no; you must learn to *live* with it." Aside from these stupid remarks of the doctors, which, if anything, only made the patients laugh at them, the patients in the pure-laugh hospitals were very happy. Everything possible was done for them to ensure their happiness. Indeed, the pure-laugh hospitals were not really hospitals at all, in the true sense of the word, but were merely isolation centers. Their only function was to prevent the inmates from infecting the outside world with their laughter-psychosis. But everything was done to make them comfortable. They could choose to work or not to work. They were given the best food, spacious living quarters, and many recreational activities. The hospitals were usually located on huge estates, and the patients were allowed to roam the beautiful grounds. All educational facilities were provided, and each hospital had a magnificent laugh-library—all the Ancient and Middle Period texts. It was also possible for the inmates to get various higher degrees of learning—indeed most of the D.H. degrees (Doctor of Humorology) were possessed by hospitalized laughers. Another wonderful thing was that laughter tended to run in families, hence entire families were incarcerated together in the pure-laugh hospitals.

Thus the conditions inside the pure-laugh hospitals were close to idyllic, except for one thing! The lives of the patients were clouded by their realization of the horrible fate of their unfortunate brethren at the laugh-scream hospitals! Good God, they said, how unfair that our brethren are screaming themselves to death while we are free to enjoy our laughter. And so every day they held religious services praying to God to relieve the sufferings of the patients at the laugh-scream hospitals. After a while, they decided that mere prayers were not enough and that perhaps there was something they could *do*. And they did indeed find something to do. More about that shortly.

The laughers in the pure-laugh hospitals held the view that *they* were sane and the rest of the world was mad. They believed that there was nothing like a sense of humor to keep a person's sanity. Yes, they held themselves to be a sane subculture living in the midst of an insane culture. Some of the more enlightened psychiatrists actually *encouraged* them in these beliefs! Although they themselves *knew* the truth, that the laughers were mad, they felt that it was psychologically good for the laughers to have the illusion that they were sane.

One of the laughers once said to a large group of his colleagues: "What a world we are living in! It is harder to cope with than the most despotic dictatorship. At least dictators are *deliberately* evil; they know deep down that they are motivated by avarice and lust for power. But these doctors at the laugh-scream hospitals! They are the maddest of all! They actually *believe* that they are helping their patients! How can one cope with *that*? Is there *nothing* we can do? Surely we can find something!"

And, as I said before, they did find something. In the first place, it occasionally happened that patients would escape from the laugh-scream hospitals, and they immediately rushed to the pure-laugh hospitals, where they were cheerfully admitted. This gave the pure-laugh patients the idea they were waiting for: They *themselves* escaped en masse from the pure-laugh hospitals, organized raids on the laugh-scream hospitals, freed all the laugh-wards, and brought all the patients back to the pure-laugh hospitals. This started the Great Decline of the laugh-scream hospitals.

Yes, indeed they declined and eventually went out of existence. Just how this happened is not fully known. It was partly the result of the raids from the pure-laugh hospitals. Another important factor was this: The laughers in the outside world all decided that it was too dangerous for them to remain where they were; they might get captured and sent to the wrong hospital. So they all voluntarily went to the pure-laugh hospitals for treatment and were all admitted. Thus came the day when no laughers were left in the outside world; t̶l̶ majority were now in the pure-laugh hospitals, and a minorit̶, in the laugh-scream hospitals.

Then came a breakdown in the morale and discipline of the laugh-scream hospitals. More and more of the doctors began getting disgusted with laughazone treatments; more and more decided that it really was *not* humane. Some of them even began to suspect that laughter was not really a sickness at all, though they dared not voice their views on pain of having their medical licenses revoked. Also, there was much outside pressure in this direction; people granted that laughter was a sickness, but felt that screaming was even worse than laughing. And so one laugh-scream hospital after another changed over to a pure-laugh hospital, until the blessed day arrived when no laugh-scream hospital was left. Now all the laughers in the world were in pure-laugh hospitals.

At this point in history, the pure-laugh hospitals got very overcrowded. Hence they spread into laugh-farms, laugh-towns, and other types of laugh-communities. The conditions in the laugh-communities were really perfect. Although the inhabitants were not free to leave, they really had everything their hearts desired. They were almost completely happy. Their only sorrow was the thought of those outside the laugh-communities who never knew the joy of laughter. What could they do about *that*? Just about nothing, they decided. But here providence intervened in a very remarkable way. What happened was this:

The outside world treated the laugh-communities with the greatest consideration. Indeed the standard of living inside the communities was far higher than outside. This had the extraordinary effect of producing a wholesale epidemic in the outside world of the insincerity psychosis. Yes, one by one the outsiders became insincere and *pretended* to be laughers in order that they might be incarcerated in the laugh-communities. These pretending laughers are not to be confused with the pseudo-laughers of the Middle Period. The pseudo-laughers

genuine quest of humor and thought that, by memoriz-
es and imitating the sound of laughter, they had ac-
a genuine sense of humor. In other words, the pseudo-
rs never tried to deceive anyone else, but they had
ghly deceived themselves. But the pretended-laughers of
resent knew perfectly well that they had no sense of
r, and they couldn't have cared less; they deliberately lied
or the purpose of joining the laugh-communities with
high standards of living. The psychiatrists in charge of
laugh-communities were completely fooled and admitted
m, but the inmates of course saw through the whole thing.
t they were happy, because they correctly foresaw what
uld happen. What happened was that the lying-laughers
ing surrounded by an enormous majority of genuine laugh-
s very soon caught the laughing sickness themselves, and in
ut a few weeks turned completely into genuine laughers. And
o one nonlaugher after another lied his way into the laugh-
communities and shortly became a genuine laugher. Then
finally even the psychiatrists succumbed, and no nonlaughers
were left in the world. The entire planet was now one huge
laugh-hospital. The Garden of Laughter had returned and
spread over the entire earth. Mankind had at last found its own
weird way of recalling the Lord and His ways. The Lord's
prophecy had come true.

Epilogue in Heaven

God lay luxuriously on His couch in Heaven, surrounded by
His choir of laughing angels. Nemod (the green animal) lay at
His feet affectionately licking His toes, and the Lord was
affectionately stroking Nemod's head. One of the angels said,
"Lord, your ways are miraculous; how *did* you do it?" The Lord
laughed and said:

"It really wasn't all that difficult! The main problem was
for me to condition Adam and Eve to believe they had free will.
These humans are really amazing; they are like children! The
only way you can get them to *do* anything is to make them

think that it is *they* who are doing it. Their pride is so gr
without having the illusion of free will, they will never g
and amount to something. So therefore I had to progra
brains so that they believed they actually had free wi
how could I do this? How could I get *any* sentient be
believe something this fantastic? The problem was not e
I had gone down and simply *told* them that they had fre
they would have been totally incapable of believing me.
would have looked at me wide-eyed and said: 'But th
fantastic! You must be *kidding*! We certainly don't *feel*
freedom!' Yes, I had been previously joking with them
much, that rather than believe such a fantastic story about
will, they would have dismissed it as another joke (and, i
way, they would have been right!). No, I was *certainly* not
one to tell them. Who then should? Well, our friend h
Nemod seemed just the one, as indeed he turned out to be
had to send them someone who seemed very serious and
little frightening. But to get Nemod to do this, I first had to
convince *him* that he had free will. How could I do this? He
obviously would not believe me if I told him; he is far too
intelligent. So I had to use something combining hypnosis and
mental telepathy. But to do this, I had to first condition *myself*
to believe *I* had free will! The reason is I had to know how it *felt*
to imagine oneself free, in order to telepathize this feeling to
Nemod. Thus I first had to program myself. This was really the
most difficult part of the entire operation! You have no idea
how hard it is for one to deliberately convince oneself of
something one knows is false just because one also knows that
this false temporary belief is useful. And I had to make sure I
would not permanently have this false belief, for if I did, I
would have been permanently insane, and hence the whole
universe would have gone insane, and the universe and I would
then have gone out of existence. So I gave myself a post-
hypnotic suggestion that the moment I had succeeded in
getting Nemod to believe that he had free will, I would
immediately regain my sanity and know again that I don't

ee will. And this indeed is what happened. Well, once I
mod to believe that he had free will, then I was able to
ulate him to think of himself as being 'evil,' 'rebellious
t me,' 'hating me,' and so forth. I made him think that I,
his creator, somehow felt 'superior' to him and was
ng it' over him. This naturally pricked his sensitivity into
ng him say: 'Who does that Lord think He is? I'll show
' In short, he *opposed* me. This was crucial for my plans.
he stole away, went down to beguile Adam and Eve. The
of the story is familiar history."

The angels laughed long and loud at the Lord's wisdom.
of them said, "And you, Nemod, when did *you* first see
ugh the Lord's game?" Nemod answered:

"Not really until the Garden of Laughter returned and took
r the planet. Well, at first, after the Lord hypnotized me into
ieving I had free will, and after He programmed me to hate
m, I indeed slunk down with very diabolical plans to corrupt
e planet. The day Adam and Eve were banished from the
rden, I knew I had won. And soon after, when the couple
ent forth to 'amount to something,' I rubbed my paws in glee!
nd then when humor started leaving the planet, how great
as my joy! My plans were working. (I had no idea at the time
that all this was really the Lord's plan!) But then when the
Middle Period came, I became gravely concerned about the
rise of the Laugh-Masters. It seemed quite possible that *they*
might be able to restore humor to the planet. It was *I* who was
responsible for the existence of the pseudo-laughers. It was I
who whispered into their souls that they could acquire a sense
of humor by memorizing jokes and training themselves to
'laugh correctly.' Yes, since humor was valued, it was essential
for me to deceive the people into *thinking* they had a sense of
humor when they really didn't.

"When the Modern Period came and people decided that
humor was something "psychotic," I was of course delighted!
And when the laugh-scream hospitals came into existence, I
was overjoyed! And when I heard all the agonized cries of the

screamers, I jumped up and down with joy! Just think
responsible for all this pain! Yes, *I* had this power!
opposed the Lord and brought all this suffering into the
Yes, little *me* has done all this! I had *really* amoun
something! But then when the laugh-scream hospitals st
to decline, I got extremely worried. What had gone wrong
my plans? Don't tell me the Lord was winning after all!
God, had I *really* amounted to something, or had I only
fooling myself? And then when the last laugh-scream hos
had disappeared, I was in a state of total panic. And whei
nonlaughers *pretended* to be laughers in order to entei
laugh-communities, and then became laughers themselv
was in total despair; I knew the game was up. I no longer h
chance. So I could only gloomily wait for the day y
laughter would totally return. And sure enough, it did!

"Then the truth of the whole situation suddenly stabb
me like a knife. I had been duped! Yes, completely and tota
duped! I suddenly saw how all my activities which I did
opposition to the Lord were merely part of the Lord's pla
Good God, I had opposed the Lord only because *He* wanted n
to! He had duped me into believing that I had free will and th
it was *I* who did all the things I had done. I had been a mer
pawn in the Lord's cosmic chess game!

"Oh, how I raged and ranted and fumed and shook my fist
at the Lord! I vowed eternal vengeance! But while raging and
ranting, I suddenly realized I was doing this only because the
Lord *wanted* me to; it was also part of the Divine Plan. In other
words, do what I would, do what I could, there was absolutely
no way I really *could* oppose the Lord; my every action was
His! Then the humor of the entire situation burst upon me! I
broke down and laughed and laughed and laughed as I had
never laughed in my life! I rolled over and over on the ground
and laughed until the tears came to my eyes! Never before had
I had such a good time and felt so deliciously *free*. Free will was
only a nightmarish delusion, and at last I was free from this
horror. And as I laughed and laughed, I became purged and

d. Evil, pride, disobedience, meanness, love of suffer-
d these things were washed away by my laughter. And
was finished laughing, I was as pure as the day I was
Now I loved the Lord, I loved the planet, I loved the
se, I loved *everything*. And so I ascended Heaven and
ced the Lord. I was nine times over the Prodigal Son."

e Lord smiled to the whole assembly and said: "Won-
are the ways of the Way. How happily all has turned out
s planet—just as I predicted."

About the Author

RAYMOND SMULLYAN is Professor of Philosophy
Indiana University. Born in New York City, he recei
his M.S. at the University of Chicago and his Ph.D.
Princeton University and has taught at Dartmou
Princeton, and the City University of New York. W
known for his many technical papers and books, he h
also published a number of popular works includi
What Is the Name of This Book? (which has been tran
lated into Italian, German, Japanese, Russian, and se
eral other languages), *The Tao Is Silent*, *The Chess My*
teries of Sherlock Holmes, and *Alice in Puzzle-Land*. A
accomplished classical pianist and professional magi
cian, he lives with his wife in Elka Park, New York.